AGRONOMY ALGORITHM

Algorithim (mathematics) helps in understanding the direct and indirect relationship of plants that exist within it and other environmental factors. This book helps to understand how yield is related to different growth parameters, how growth is influenced by different environmental phenomenon, how best the resources can be used for crop production etc. The numerical examples in the book guide a student to coordinate the different parameters and understand the subject of Agronomy well. This book is divided into thirteen chapters and covers comprehensively the different agronomic aspects to understand the science of mathematical Agronomy to meet the current and future challenges related to cropping practices.

Dr. Neetu Sharma presently working as Associate Professor, Division of Agronomy, Faculty of Agriculture, SKUAST-J, Chatha. She has experience of more than 14 years of experience in teaching, research and extension.

Dr. B.C. Sharma presently working as Professor and Head, Division of Agronomy, Faculty of Agriculture, SKUAST-J, Chatha. And a recipient of best teacher award and innovative team award for his outstanding contribut ons in Agronomy and has about 20 years of teaching experience out of 37 years of his active service.

Dr. Anil Kumar, presently working as Professor Division of Agronomy of Sher-e-Kashmir University of Agricultural Sciences and Technology of Jammu (SKUAST-J) an ISWS Fellow graduated in Agriculture form the SKUAST-J&K, obtained his M.Sc. in Agronomy from HPKV-Palampur and Ph.D. from PAU-Ludhiana.

Dr. Rakesh Kumar presently working as Assistant Professor, Division of Agronomy, Faculty of Agriculture, SKUAST-J, Chatha. He has experience of more than 8 years of experience in teaching, research and extension. He has to her credit more than 15 research publications in national and international journals, 8 technical bulletin, 2 practical manual and 20 abstracts.

AGRONOMY ALGORITHM

**Neetu Sharma, B.C. Sharma, Anil Kumar
and Rakesh Kumar**

CRC Press
Taylor & Francis Group
Boca Raton London New York

CRC Press is an imprint of the
Taylor & Francis Group, an **informa** business

NARENDRA PUBLISHING HOUSE
DELHI (INDIA)

First published 2023
by CRC Press
4 Park Square, Milton Park, Abingdon, Oxon, OX14 4RN

and by CRC Press
6000 Broken Sound Parkway NW, Suite 300, Boca Raton, FL 33487-2742

CRC Press is an imprint of Informa UK Limited

British Library Cataloguing-in-Publication Data
A catalogue record for this book is available from the British Library

ISBN: 9781032388847 (hbk)
ISBN: 9781032388854 (pbk)
ISBN: 9781003347286 (ebk)

DOI: 10.4324/9781003347286

Typeset in Times New Roman, Arial, Symbol, Calibiri and Newcentury
by Amrit Graphics, Delhi 110032

Contents

Preface

There is no existence of Agronomy without numerical problem. The Agronomy subject deals with the principles and practices of soil, water and crop management which involves a lot of numerical problem. Unless and until, the mathematical relationship that exists among different soil,water, atmosphere and crop management practices are understood, the knowledge of Agronomy is not complete.

There are no comprehensive textbooks covering the numerical component of Agronomy. Undergraduate and postgraduate students have to refer various books to understand the numerical. However, in most of the cases various crop management indices are not explained with numerical examples. As such students memorize the formula or indices without clear understanding of the logic behind them.

This book will serve as a textbook for undergraduate and postgraduate students of agriculture. This textbook is written to help the students to understand the numerical involved in Agronomy. The formulae and indices commonly used in various textbooks and agronomic journals are discussed with numerical examples. The authors does not claim perfection but all the possible efforts have been made to make this book useful to the agriculture students. A number of exercises along with answers have been given at the end of chapter.

The authors will be grateful to the teachers and students if they point out the error and send suggestions and constructive criticism for future revision and improvement of the book.

Authors

INTRODUCTION

Numerical is the heart of agronomy. The subject Agronomy deals with principles and practices of soil, water and crop management for higher and sustainable production. As such, there are direct relationships between plant population and growth, yield attributing characters and yield. With different row arrangement of crops, plant stand varies. Similarly nutrient requirement varies from crop to crop, sometimes, from variety to variety. Even, depending upon the source, fertilizer requirements vary for same amount of nutrients. Thus, precise cultivation is required to get the maximum benefit of applied inputs.

Efficiency of nutrient is of major concern now days. Various efficiency parameters have suggested by various researchers to judge the efficiency of applied nutrients. Similarly, without the knowledge of mathematics, water management is hardly understandable to the students. The volume and mass relationship of soil, efficiency of water application, irrigation requirement of crops etc. require systematic calculations. The success of chemical weed management depends upon accurate application of herbicides. Overdose will kill the crop and underdose will fail to control the weeds. Likewise efficiencies of weed management practices (including herbicides) are judged by various indices which require elaborate numerical exercises for better understanding of the indices.

Assessment of land use and yield advantage helps in adoption of proper crop sequence as well as proper crop mixture in suitable proportion. Growth and development study is required to understand the effect of various management practices on crop growth. The growth and development are studied by various indices which help us to understand the effect of different management practices and performance of genotypes. Economics is also an important study of agronomic research. The viability of new agronomic technology is judged by various economic indices.

Thus, the numerical involved in all the branches of Agronomy and various universally accepted indices to judge the efficiency of crop management practices are presented with practical examples.

PLANT POPULATION

Number of plants in a particular area depends on the canopy coverage of the individual plant. If the vigor of the plant is less, canopy coverage is less and requirement of plants per unit area will be more and vice-versa. For example, if the canopy coverage of individual plant of Species A and B are 0.04 m^2 and 0.25 m^2, number of plants/ m^2 will be 25 and 4, respectively. Thus, spacing is maintained in such a way that its canopy mathematically covers the entire area to intercept maximum sunshine without interfering the neighboring crop plants. As such, a plant spacing 20 cm x 20 cm means that individual plant of the crop will occupy 400 cm^2 or 0.04 m^2

2.1 PLANT POPULATION IN UNIFORM PLANT SPACING

Example 1: What are the canopy coverage of a plant of crop species A and B, if the spacing are (i) 20 cm × 20 cm (ii) 15 cm × 15 cm, respectively.

Solution

(i)   Canopy coverage of a plant of Species A = 20 × 20 cm^2 = 400 cm^2

$$= \frac{400}{100 \times 100} m^2 = \mathbf{0.04\ m^2}$$

(ii) Canopy coverage of a plant of Species A = 15 × 15 cm^2 = 225 cm^2

$$= \frac{225}{100 \times 100} m^2 = \mathbf{0.0225 \ m^2}$$

Example 2: Calculate the plant population/m^2, if the spacings of two crop species A and B are (i) 20 cm × 20 cm (ii) 15 cm × 15 cm, respectively.

Solution

(i) Canopy coverage of a plant of Species A = 20 × 20 cm^2 = 400 cm^2

$$= \frac{400}{100 \times 100} m^2 = 0.04 \ m^2$$

$$\text{Plant Poulation/m}^2 = \frac{1}{\text{Row to row distance (m)} \times \text{Plant to plant distance (m)}}$$

i.e. 0.04 m^2 is covered by = 1 plant

$$1 \ m^2 \text{ is covered by} = \frac{1}{0.04} = \mathbf{25 \ plants}$$

(ii) Canopy coverage of a plant of Species A= 15 × 15 cm^2 = 225 cm^2

$$= \frac{225}{100 \times 100} m^2 = 0.0225 \ m^2$$

i.e. 0.0225 m^2 is covered by = 1 plant

$$1 \ m^2 \text{ is covered by} = \frac{1}{0.00225} = \mathbf{44.44 \ plants}$$

Example 3: Calculate the plant population/ha, if the spacings of two crop species A and B are (i) 20 cm × 20 cm (ii) 15 cm × 15 cm, respectively.

Solution :

(i) Plant population/m^2 of Species A= $\dfrac{1}{0.20 \times 0.20} = \mathbf{25}$

$$\text{Plant Poulation/ha} = \frac{10,000}{\text{Row to row distance (m)} \times \text{Plant to plant distance (m)}}$$

Therefore. Plant population/ha = $25 \times 10,000 = \textbf{2,50,000}$

(ii) Plant population/m^2 of Species B = $\dfrac{10,000}{0.15 \times 0.15} = 4,44,444.44$

Therefore. Plant population/ha = **4,44,444.44**

Example 4: Calculate the number of seedlings required/ha if two rice variety A and B are transplanted at the spacing (i) 20 cm x 20 cm (ii) 15 cm x 15 cm,respectively with 2 seedlings/hill

Solution:

(i) Number of hills/m^2 for variety A = $\dfrac{1}{0.20 \times 0.20} = 25$

  Therefore, number of seedlings required/m^2 = $25 \times 2 = 50$

  Seedlings required/ha = $50 \times 10,000 = 5,00,000$

(ii) Seedlings required/ha = $\dfrac{10,000 \times 2}{0.15 \times 0.15} = 8,88,888.88 = \textbf{8,88,889}$

Example 5: Calculate the number of seedlings required to transplant 1,000 m^2 with a spacing of 20 cm x 15 cm with 3 seedlings/hill,if 10% seedlings are damaged after transplanting.

Solution:

(i) Number of seedlings required for 1000m^2 = $\dfrac{1000 \times 3}{0.20 \times 0.15} = 1,00,000$

  Seedlings damaged during transplanting = $\dfrac{1,00,000 \times 10}{100} = 10,000$

Therefore, total seedling required = **1,10,000**

Example 6: Calculate the plant population/ha if maize is dibbled at a spacing of (i) 75 cm x 25 cm (ii) 60 cm x 25 cm

Solution:

(i)   Plant population/ha $= \dfrac{10,000}{0.75 \times 0.25} = \mathbf{53,333}$

(ii) Plant population/ha $= \dfrac{10,000}{0.60 \times 0.25} = \mathbf{66,667}$

Example 7: A student recorded the plant population of wheat and moong in running meter. If the populations were 40/m and 11/m, respectively. Calculate the plant population/m^2. The row to row spacing for wheat and moong were 20 cm and 30 cm, respectively .

Solution:

Plant to plant distance of wheat $= \dfrac{100}{40} = 2.50$ cm

Plant to plant distance of moong $= \dfrac{100}{11} = 9.09$ cm

Therefore, spacing of wheat= 20 cm $\times$ 2.50

and spacing of moong= 30 cm $\times$ 9.09 cm

Plant population of wheat/m^2 $= \dfrac{1}{0.20 \times 0.025} = \mathbf{200\ plants}$

Plant population of moong/m^2 $= \dfrac{1}{0.30 \times 0.0909} = \mathbf{36.67\ plants}$

2.2 PLANT POPULATION IN PAIRED ROW

In paired row planting, two adjacent rows of bases crop are paired. Thus, spaces between the successive closer rows (pair rows) are widen and these spaces may be utilized for planting intercrop. However,plant population of the main crop is kept constant. For Example, maize crop is sown at a spacing of 75 cm x 25 cm. It means that all the plants are equally spaced in line at 75 cm interval (Fig.1 a) and within the row, plants are uniformly spaced at 25 cm interval. If pairwise, two rows are brought closer by 15 cm (7.5 cm from both sides), the inter row space will be 90 cm between two pair rows (Fig.1b). Thus, the spacing will be 90 cm x 60 cm x 25 cm, which will be equivalent to 75 cm x 25 cm in terms of plant population.

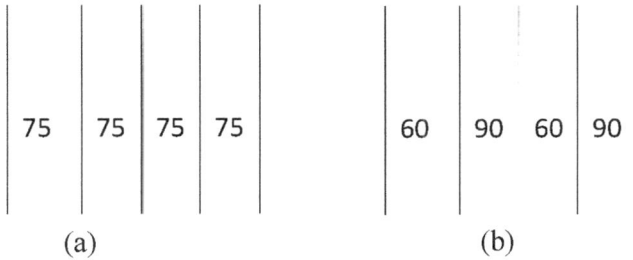

(a)   (b)

Fig. 1. **(a) Rows are equally spaced at 75 cm (b) Paired row planting at 90 cm and 60 cm**

$$\frac{\text{Plant Population} = 10,0000}{\text{Mean paired row and inter row distance (m)} \times \text{Plant to plant distance (m)}}$$

Example 8: Calculate the plant population of maize in 1 ha. Area, if the crop is raised in paired row system at a spacing of 90 cm x 60 cm x 25 cm.

Solution:

$$\text{Plant population/ha} = \frac{10,000}{\frac{0.90 + 0.60}{2} \times 0.25} = \frac{10,000}{0.75 \times 0.25}$$

$$= \mathbf{53,333 \ plants}$$

Example 9 : (i) if 1 row of greengram with 10 cm plant to plant spacing (ii) two rows of greengram with 30 cm x 10 cm spacing are sown in the 90 cm inter space of example 8. Calculate the plant population in each case.

Solution:

(i) Here, spacing between two rows of greengram = 45+60+45=150 cm

$$\text{Therefore, plant population/ha} = \frac{10,000}{1.50 \times 0.10} = 66,666.6$$

$$= \textbf{66,667 plants}$$

(ii) Should we multiply the result obtained in (i) Yes

The result will be = 66,666.6 × 2= 1,33,333

Can we also not think like that greengram is sown in paired row of 120 cm x 30 cm x 10 cm and maize is dibbled in between this?

$$\text{Thus, plant population/ha} = \frac{10,000}{\dfrac{120 + 130}{2} \times 0.10}$$

$$= \textbf{1,33,333 plants}$$

EXERCISES

1. What is the canopy coverage of individual plant, if the plants are spaced at:

    (i) 10 cm × 10 cm   (ii) 10 cm × 5 cm

    (iii) 20 cm × 20 cm   (iv) 20 cm × 15 cm

    (v) 50 cm × 20 cm   (vi) 50 cm × 5 cm

    (vii) 50 cm × 15 cm   (viii) 50 cm × 10 cm

2. Calculate the plant stand /m^2 for the spacing given in Example 1

3. Calculate the plant stand /ha for the spacing given in Example 1

4. If the crop is planted at a spacing of 50 cm × 50 cm, calculate the plant stand

    a) per acre   b) per hectare

    c) per 500 m^2

5. Calculate the number of seedlings/ha if a rice variety is transplanted at 20 cm × 15 cm spacing with

    (i) 2 seedlings/hill   (ii) 3 seedlings/hill

6. Calculate the number of seedlings to be transplanted in one acre with a spacing of 20 cm × 20 cm with 2 seedlings/hill, if the 10% seedlings are damaged after transplanting.

⋮ ANSWERS ▶

1. (i)   0.01m^2 (ii)   0.005 m^2

    (iii) 0.04 m^2 (iv) 0.03 m^2

    (v)   0.1 m^2 (vi) 0.025 m^2

    (vii) 0.075 m^2 (viii) 0.05 m^2

2. (i)   100   (ii)   200

    (iii) 25   (iv) 33

    (v)   10   (vi) 40

    (vii) 13   (vii) 20

3. (i)   10,00,000   (ii)   20, 00,000

    (iii) 2,50,000   (iv) 3,33,333

    (v)   1,00,000   (vi) 4,00,000

    (vii) 1,33,333   (vi) 2,00,000

4. (i)   5,333   (ii)   16,000

    (iii) 2,000

5. (i)   6,66,667   (ii)   10,00,000

6. 2,20,000

SEED REQUIREMENT

If the plant population per unit area is worked out, requirement of seeds can easily be calculated, provided the weight of seed and its germination percentage is known. Let us consider the example plant population/ha is 2,50,000. If the weight of single seed is 0.02 g and germination is 100%,weight of seed required/ha will be 2,50,000 × 0.02 g= 5,000g= 5 kg. If the germination is less than 100 %,proportionately higher amount of seeds will have to be sown to desired population. Under 80% germination,100/80=1.25 times higher seeds must be sown. Thus, seed requirement will be 2,50,000 × 0.02 × (100/80) g= 6250 g = 6.25 kg. Likewise, if the seed is not 100% pure, proportionately higher amount of seeds will have to be sown to maintain the desired population. As such, if the purity of seed is 95 %, 100/95=1.0526 times higher seeds than calculated above will have to be sown. Thus, seed requirement will be 2,50,000 × 0.02 × (100/80) × (100/95) g = 6579 g = 6.579 kg

$$\text{Seed requirement (g/m}^2) = \frac{\text{Area (m}^2) \times \text{Weight of seeds (g)} \times 100 \times 100}{\text{Spacing (m}^2) \times \text{Germination p.c.} \times \text{Purity p.c}}$$

Seed requirement (kg/ha) =

$$= \frac{\text{Area (ha.)} \times \text{Test weight (g)} \times 100 \times 100}{1,000 \times \text{Spacing (m}^2) \times 1000 \times \text{Germination p.c.} \times \text{purity p.c.}}$$

Example 1: Find out the seed requirement of blackgram/m^2 from the following observation:

(a) Spacing: 30 cm × 10 cm

(b) Test weight: 50 g

(c) Germination per centage:90

Solution:

$$\text{Seed requirement (g/m}^2) = \frac{\text{Area (m}^2) \times \text{Weight of seeds (g)} \times 100}{\text{Spacing (m}^2) \times \text{Germionation p.c.}}$$

$$= \frac{1 \times 50 \times 100 \text{ g}}{0.30 \times 0.10 \times 1000 \times 90}$$

$$= \textbf{1.85 g}$$

Example 2: Find out the seed requirement of blackgram/m^2 from the following observation:

(a) Spacing: 30 cm × 10 cm

(b) Test weight: 50 g

(c) Germination per centage: 90

(d) Purity per centage: 95

Solution:

$$\text{Seed requirement (g/m}^2) = \frac{\text{Area (m}^2) \times \text{Weight of seeds (g)} \times 100}{\text{Spacing (m}^2) \times \text{Germination p.c.} \times \text{purity p.c.}}$$

$$= \frac{\text{Area (m}^2) \times \text{Weight of seeds (g)} \times 100}{0.3 \times 0.1 \times 1,000 \times 90 \times 95} \text{ g} = \textbf{1.95 g}$$

Example 3: Find out the seed requirement of blackgram for 1000m² from the following observation:

(a) Spacing: 30 cm x 10 cm

(b) Test weight: 50 g

(c) Germination per centage:90

(d) Purity per centage: 95

Solution:

Seed requirement (g/m²)

$$= \frac{\text{Area (m}^2) \times \text{Test weight of seeds (g)} \times 100 \times 100}{1{,}000 \times \text{Spacing (m}^2) \times \text{Germination p.c.} \times \text{purity p.c.}}$$

$$= \frac{1000 \times 50 \times 100 \times 100}{0.3 \times 0.1 \times 1000 \times 90 \times 95} = \textbf{1949 g}$$

Example 4: Find out the seed requirement of blackgram for 1 hectare from the following observation:

(a) Spacing: 30 cm x 10 cm

(b) Test weight: 50 g

(c) Germination per centage:90

(d) Purity per centage: 95

Solution:

Seed requirement for 1 ha.

$$= \frac{\text{Area (m}^2) \times \text{Test weight (g)} \times 100 \times 100}{1{,}000 \times \text{Spacing (m}^2) \times 1{,}000 \times \text{Germination p.c.} \times \text{purity p.c.}}$$

$$= \frac{10{,}000 \times 50 \times 100 \times 100}{1000 \times 0.3 \times 0.1 \times 1{,}000 \times 90 \times 95} = \textbf{19.49 kg}$$

Example 5: Find out the seed requirement of blackgram for 100 hectare area from the following observation:

(a) Spacing: 30 cm x 10 cm

(b) Test weight: 50 g

(c) Germination percentage:90

(d) Purity percentage: 95

Solution:

Seed requirement for 100 ha.

$$= \frac{Area\ (m^2) \times Test\ weight\ (g) \times 100 \times 100}{1,000 \times Spacing\ (m^2) \times 1,000 \times Germination\ p.c. \times purity\ p.c.}$$

$$= \frac{10,000 \times 100 \times 50 \times 100 \times 100}{1000 \times 0.3 \times 0.1 \times 1,000 \times 90 \times 95} = \textbf{19.49 kg}$$

Example 6: Find out the seed requirement of maize for 5 hectare area from the following observation:

(a) Spacing: 75cm × 30 cm

(b) Test weight: 250 g

(c) Germination percentage: 85

(d) Purity percentage: 98

Solution:

Seed requirement for 100 ha.

$$= \frac{10,000 \times Area\ (ha) \times Test\ weight\ (g) \times 100 \times 100}{1,000 \times Spacing\ (m^2) \times 1,000 \times Germination\ p.c. \times purity\ p.c.}$$

$$= \frac{10,000 \times 5 \times 250 \times 100 \times 100}{1000 \times 0.75 \times 0.30 \times 1,000 \times 85 \times 98} = \textbf{66.69 kg}$$

Example 7: Find out the seed requirement of maize for 5 hectare area from the Example 6, if 10% extra seed by weight is required for gap filling.

Solution:

Seed required for gap filling = $66.69 \times \dfrac{10}{100} = 6.67$ kg

Therefore, total seed required for gap filling= 66.69 + 6.67= **73.36 kg**

Example 8: Find out the seed requirement of upland direct seeded rice for sowing 1 hectare area from the following observation:

(a) Spacing: 20cm × 3 cm

(b) Test weight: 24 g

(c) Germination per centage:85

(d) Purity per centage: 98

(e) Seed damaged by birds: 5% for sown seed

Solution:

Seed required (without considering bird damage)

$$= \frac{10,000 \times \text{Area (ha.)} \times \text{Test weight(g)} \times 100 \times 100}{1,000 \times \text{Spacing (m2)} \times 1,000 \times \text{Germination p.c} \times \text{purity p.c.}}$$

$$= \frac{10,000 \times 1 \times 24 \times 100 \times 100}{1,000 \times 0.20 \times 0.03 \times 1,000 \times 85 \times 98} = \textbf{48.02 kg}$$

Therefore, seed required for gap filling $=48.02 \times = \dfrac{5}{100} = \textbf{2.40 kg}$

Therefore, total seed requirement= 48.02 kg + 2.40 kg= **50.42 kg**

Example 9: Find out the seed requirement of upland direct seeded rice for sowing one hectare area from the following observation:

(a) Spacing: 20cm × 3 cm

(b) Test weight: 24 g

(c) Germination per centage: 85

(d) Purity per centage: 98

(e) Seed damaged by birds: 5% for sown seed

Solution:

Seed required (without considering bird damage)

$$= \frac{10{,}000 \times \text{Area (ha.)} \times \text{Test weight(g)} \times 100 \times 100}{1{,}000 \times \text{Spacing (m}^2) \times 1{,}000 \times \text{Germination p.c} \times \text{purity p.c.}}$$

$$= \frac{10{,}000 \times 1 \times 24 \times 100 \times 100}{1{,}000 \times 0.20 \times 0.03 \times 1{,}000 \times 85 \times 98} = 48.02 \text{ kg}$$

Therefore , 5% of sown seeds are damaged by birds, seed required to be sown =

$$48.02 \times \frac{100}{95} = \textbf{50.55 kg}$$

Example 10: Find out the seed requirement of jute in kg/ha from the following observation:

(a) Spacing: 25 cm × 10 cm

(b) Test weight: 6 g

(c) Germination percentage: 95

(d) Purity percentage: 98

Solution:

Seed required (kg/ha)

$$= \frac{10{,}000 \times \text{Area (ha.)} \times \text{Test weight(g)} \times 100 \times 100}{1{,}000 \times \text{Spacing (m2)} \times 1{,}000 \times \text{Germination p.c} \times \text{purity P.c.}}$$

$$= \frac{10{,}000 \times 1 \times 6 \times 100 \times 100}{1{,}000 \times 0.25 \times 0.10 \times 1{,}000 \times 95 \times 98} = 2.58 \text{ kg}$$

Therefore, seed rate=**2.58 kg/ha**

Example 11: Find out the seed rate of rice from the following observation:

(a) Spacing: 20 cm × 20 cm

(b) Test weight:16 g

(c) Germination percentage: 95

(d) Purity percentage: 98

(e) No. of seedlings/hill: 2-3

(f) No. of damaged seedlings: 10% of uprooted seedlings

(g) Seedlings required for gap filling: 5% of transplanted seedlings

Solution:

$$\text{Seedlings required for transplanting/ha} = \frac{10,000 \times \text{No. of seedlings/hill}}{\text{Spacing}}$$

$$= \frac{10,000 \times 2.5}{0.20 \times 0.20} = 6,25,000$$

(Since 2-3 seedlings/hill are to be transplanted, the average value is considered)

$$\text{Seedlings required for gap filling} = 6,25,000 \times \frac{5}{100} = 31,250$$

Thus, total seedlings required for transplanting and gap filling

$$= 6,25,000 + 31,250 = 6,56,250$$

Since,10% seedlings are damaged during uprooting,seedlings need to be uprooted for transplanting and gap filling $= 6,56,250 \times 100/90 = 7,29,167$

Therefore, **seed required/ha**

$$= \frac{\text{Area (ha)} \times \text{Test weight} \times 100 \times 100 \times \text{seedlings needed/ha}}{1,000 \times 1,000 \times \text{Germination percentage} \times \text{Purity percentage}}$$

$$= \frac{1 \times 16 \times 100 \times 100 \times 7,29,167}{1000 \times 1000 \times 95 \times 98} = 12.53 \text{ kg}$$

Therefore, seed rate = **12.53 kg/ha.**

Example 12: Find out the seed rate of rice from the following observation:

(a) Spacing: 20 cm × 20 cm

(b) Test weight: 16 g

(c) Germination percentage: 95

(d) Purity percentage: 98

(e) No. of seedlings/hill: 2-3

(f) No. of damaged seedlings: 10% of uprooted seedlings

(g) Seedlings required for gap filling: 5% of transplanted seedlings

(h) Seed damaged by birds: 20% of sown seeds

Solution:

Seedlings required for transplanting/ha

$$= \frac{10,000 \times \text{No. of seedlings/hill}}{\text{Spacing}}$$

$$= \frac{10,000 \times 2.5}{0.20 \times 0.20} = 6,25,000$$

(Since 2-3 seedlings/hill are to be transplanted, the average value is considered)

Seedlings required for gap filling $= 6,25,000 \times \dfrac{5}{100} = 31,250$

Thus, total seedlings required for transplanting and gap filling

$$= 6,25,000 + 31,250 = 6,56,250$$

Since, 10% seedlings are damaged during uprooting, seedlings need to be uprooted for transplanting and gap filling $= 6,56,250 \times 100/90 = 7,29,167$

Again 20% seeds are damaged by birds, seeds need to be sown for transplanting and gap filling

$$= 7,29,156 \times \frac{100}{80} = 9,11,459$$

Therefore, **seed required/ha**

$$= \frac{\text{Area (ha)} \times \text{Test weight} \times 100 \times 100 \times \text{seedlings needed/ha}}{1,000 \times 1,000 \times \text{Germination percentage} \times \text{Purity percentage}}$$

$$= \frac{1 \times 16 \times 100 \times 100 \times 9,11,459}{1000 \times 1000 \times 95 \times 98} = 15.66 \text{ kg}$$

Therefore, seed rate = **15.66 kg/ha.**

Example 13: Calculate the seed rate of groundnut in terms of kernels and unshelled pods from the following observations:

(a) Spacing: 35 cm × 10 cm

(b) Germination percentage: 90

(c) Purity percentage: 95

(d) Test weight of kernels: 370 g

(e) Shelling percentage: 65

Solution:

Seed rate of kernels/ha

$$= \frac{10,000 \times \text{Area (m}^2) \times \text{Test weight (g)} \times 100 \times 100}{\text{Spacing} \times 1000 \times 1000 \times \text{Germination p.c.} \times \text{Purity p.c.}}$$

$$= \frac{10,000 \times 370 \times 100 \times 100}{0.35 \times 0.10 \times 1000 \times 1000 \, 90 \times 95} \text{ kg}$$

$$= 123.64 \text{ kg}$$

Seed rate of unshelled pods/ha $= \dfrac{\text{Amount of kernel (kg/ha)} \times 100}{\text{Shelling p.c.}}$

$$= 123.6 \times \frac{100}{65} = \textbf{190.22 kg/ha}$$

Example 14: Calculate the weight of potato tubers required for 5,000 m^2 from the following details:

(a) Spacing: 50 cm × 20 cm

(b) Average weight of tuber : 25 g

Solution:

$$\textbf{Potato tuber requirement} = \frac{\text{Area (m}^2) \times \text{ Tuber weight (g)}}{\text{Spacing} \times 1000 \times 1000}$$

$$= \frac{5000 \times 25}{0.5 \times 0.2 \times 1000} \text{ kg} = 1{,}250 \text{ kg} = \textbf{12.50 q}$$

Example15: If the rotting loss during transportation is 5%,calculate the weight of tubers to be procured in Example 14

Solution:

$$\text{Potato tubers to be procured} = \frac{12.50 \times 1000}{95} = \textbf{13.16 q}$$

Example 16: If the rotting in the field after transplanting is 5 % which are to be planted again (gap filling),calculate the weight of tubers to be procured in Example 14

Solution:

Potato tuber required for planting = 12.50 q

$$\text{Tuber required for gap filling} = 12.50 \times \frac{5}{100} \text{ q} = 0.63 \text{ q}$$

Therefore potato tubers to be procured = 12.50 + 0.63 = **13.13 q**

Example 17: If the rotting in the field after transplanting is 5 % and rotting in the field after transplanting is 5 % which are to be planted again (gap filling),calculate the weight of tubers to be procured in Example 14

Solution:

Potato tuber required for planting = 12.50 q

Tuber required for gap filling $= 12.50 \times \dfrac{5}{100}$ q $= 0.63$ q

Therefore potato tubers required for planting and gap filling

$$= 12.50 + 0.63 = 13.13 \text{ q}$$

Potato tubers to be procured $= \dfrac{13.13 \times 100}{95} = \mathbf{13.82 \text{ q}}$

Example 18: Find out the weight of potato tubers to be procured for 1 ha. area from the following details:

(a) Average weight of tuber: 90 g

(b) Number of pieces/tuber:3

(c) Spacing: 50 cm × 20 cm

Solution:

Number of tuber pieces required/ha $= \dfrac{10,000}{0.50 \times 0.20} = 1,00,000$

Therefore, each tuber will give $= 3$ pieces

Therefore, number of tubers required to get 1,00,000 pieces

$$= \dfrac{1,00,000}{3} = 33,333.3$$

Weight of each tuber $= 90$ g

Weight of tuber required for planting $= 33,333.3 \times 90 = 30,00,000$ g

$$= 3,000 \text{ kg} = \mathbf{30 \text{ q}}$$

Example 19: Calculate the seed rate of sugarcane from the following details:

(a) System of planting: End to end planting in furrows in horizontal position

(b) Type of planting materials: 3 budded setts

(c) Length of setts: 25 cm

(d) Number of buds/sugarcane: 36

(e) Average number of damaged bud in each cane : 2

(f) Average weight of canes: 1 kg

(g) Row to row distance: 90 cm

Solution:

Total number of setts required/ha = $\dfrac{10,000}{0.90 \times 0.25}$ = 44,444.44

Number of buds required/ha.= 44,444.44 × 3 = 1,33,333.32

Number of effective buds/cane = 36-2 = 34

Number of canes required/ha.= $\dfrac{1,33,333.32}{34}$ = 3,921.57

Weight of cane required/ha = 3,921.57 × 1 = 39,22 kg = 39.22 q

Therefore, seed rate = **39.22 q/ha**

EXERCISES

1. Find out the seed requirement of cotton seed for 2 ha. of land from the following particulars:

   (a) Spacing : 75 cm × 50 cm

   (b) Test weight of fuzzed cotton: 150 g

   (c) Test weight of unfuzzed cotton seed: 160 g

   (d) Germination per centage of seed : 85

   (e) Purity percentage of seed: 98

2. Find out the seed requirement of upland rice for 0.5 hectare area from the following particulars:

   (a) Spacing: 20 cm × 3 cm

   (b) Test weight of seed: 23 g

   (c) Germination percentage of seed: 85

   (d) Purity percentage of seed: 98

   (e) Seed required for gap filling: 5% by weight of seed required for sowing

3. Find out the seed requirement of upland rice for 0.5 hectare area from the following particulars:

   (a) Spacing: 20 cm × 3 cm

   (b) Test weight of seed: 23 g

   (c) Germination percentage of seed:85

   (d) Purity percentage of seed:98

   (e) Number of damaged seeds by birds which need to be replaced by another seed : 10 seeds/m^2

4. Find out the seed requirement of lowland rice for 1.0 hectare area from the following particulars:

   (a) Spacing: 20 cm × 15 cm

   (b) Test weight of seed: 18

   (c) Germination percentage of seed: 85

   (d) Purity percentage of seed: 98

   (e) Number of damaged seeds at the time of uprooting (as percent of uprooted seedlings): 10

   (f) Number of seedlings required for gap filling:3/m^2

5. Find out the seed requirement of greengram for 20 hectare area from the following observation:

   (a) Spacing: 30 cm × 10 cm

   (b) Test weight: 45 g

   (c) Germination percentage: 85

   (d) Purity percentage: 98

6. Find out the seed requirement of maize for 10 hectare area from the following observation:

   (a) Spacing: 75 cm × 30 cm

   (b) Test weight:250 g

   (c) Germination percentage: 90

   (d) Purity percentage:95

7. Find out the seed requirement of maize for 10 hectare area from the Example 6, if 15% extra seed by weight is required for gap filling.

8. Find out the seed requirement of upland direct seeded rice for sowing 1 hectare area from the following observation:

   (a) Spacing: 20 cm × 3- 5 cm

   (b) Test weight: 20 g

   (c) Germination percentage:90

   (d) Purity percentage: 95

   (e) Seed required for gap filling: 20 % by weight of seeds required for sowing

9. Find out the seed requirement of upland direct seeded rice for sowing 1 hectare area from the following observation:

   (a) Spacing: 20 cm × 3- 5 cm

   (b) Test weight: 20 g

   (c) Germination percentage:90

   (d) Purity percentage: 95

   (e) Seed damaged by birds: 20 % for sown seed

10. Find out the seed rate of *kharif* rice from the following observation:

    (a) Spacing: 20 cm × 20 cm

    (b) Test weight:16 g

    (c) Germination percentage:95

    (d) Purity percentage: 98

    (e) Number of seedlings per hill:2-3

    (f) Number of damaged seedlings: 10% of uprooted seedlings

    (g) Seedlings required for gap filling: 2/ m^2

11. Calculate the weight of potato tubers required for 1 hectare area from the following details:

    (a) Spacing: 50 cm × 20 cm

    (b) Average weight of tuber: 25 g

    (c) Rotting loss:10%

12. Find out the weight of tubers to be procured for 1 hecatre area from the following details:

   (a) Average weight of tuber:90 g

   (b) Number of pieces/tuber:3

   (c) Spacing: 50 cm × 20 cm

   (d) Rotting loss during transportation: 5%

ANSWERS

1. Fuzzed seed: 9.60 kg Unfuzzed seed: 10.24 kg

2.	24.2 kg	3.	48.8 kg	4.	24.0 kg
5.	360.1 kg	6.	130 kg	7.	149.5 kg
8.	35.1 kg	9.	36.5 kg	10.	12.3 kg
11.	27.78 q	12.	34.74 q		

EXPECTED YIELD

Yield of a crop is the function of various yield attributing characters. Higher the value of the yield attributes, maximum is the yield. As for example, in case of rice, higher is the number of effective tillers, maximum will be the yield provided other factors (test weight, number of grains/panicle, number of hills/area) remain constant. Likewise, in crops like blackgram,greengram, mustard and rapeseed, yield will be influenced by number of branches/plant, number of pods or siliquae/branch, number of seeds/pod or siliquae and test weight.

Example 1: Calculate the rice grain yield/m^2, if grain weight/panicle and number of panicles/ m^2 is 3 g and 200, respectively.

Solution:

Grain weight/panicle= 3 g

Grain weight of 200 panicles = 3 × 200g = 600 g

Therefore,rice grain yield/m^2 = **600 g**

Example 2: Calculate the rice grain yield in q/ha ,if grain weight/panicle and number of panicles/m^2 is 3 g and 200,respectively

Solution:

Grain weight/panicle = 3 g

Grain weight of 200 panicles = 3 × 200g = 600 g

Therefore, rice grain yield/m^2 = 600 g

Rice grain yield/ha = 600 × 10,000 (Since 1 ha. = 10,000 m^2)

$$= 60, 00,000 \text{ g} = 6,000 \text{ kg} = 60.00 \text{ q/ha}$$

Rice grain yield = **60.00 q/ha**

Example 3: Estimate the yield of sugarcane,cane juice,sugar and jaggri in tons/ha from the following observations:

(a) Average number of millable canes/clump: 15

(b) Average weight of canes: 600 g

(c) Spacing of clumps: 90 cm × 60 cm

(d) Per cent extractable juice in canes: 70

(e) Per cent sugar in juice: 16

(f) Per cent jaggari in juice: 20

Solution:

Yield of Sugarcane in t/ha

$$= \frac{\text{Area} \times \text{Average weight of canes} \times \text{No. of millable canes/clump}}{\text{Spacing} \times 100 \times 100 \times 10}$$

Yield of Sugarcane in t/ha$= \dfrac{10,000 \times 600 \times 15}{0.9 \times 0.6 \times 1000 \times 100 \times 10} =$ **166.67t/ha.**

Total amount of expected juice in canes in t/ha.

$$= \frac{166.57 \times 70}{100} = \textbf{116.67 t/ha.}$$

Total amount of expected cane sugar in t/ha $= 116.67 \times \dfrac{16}{100} =$ **18.67 t/ha**

Total amount of expected cane jaggari in t/ha.$= 116.67 \times \dfrac{20}{100} =$ **23.33 t/ha**

Example 4: Estimate the tuber yield of potato in q/ha from the data mentioned below:

(a) Spacing: 60 cm × 25 cm

(b) Average number of tubers/plant: 14

(c) Average weight of tubers = 35 g

Solution :

$$\text{Yield} = \frac{\text{Area} \times \text{Average no. of tubers/plant} \times \text{Average weight of tubers}}{\text{Spacing} \times 1000 \times 100}$$

$$\text{Yield} = \frac{10000 \times 14 \times 35}{0.6 \times 0.25 \times 1000 \times 1000 \times 100} = \textbf{326.67 q/ha}$$

Example 5: Find out the yield (t/ha) of cotton in terms of seed cotton, cotton seed, lint, ginning percentage from the data mentioned below:

(a) Plant Spacing : 90 cm × 60 cm

(b) Average number of sympodial branches/plant:5

(c) Average number of bolls/branch: 4

(d) Number of locules/boll : 4

(e) Average number of seeds/locule: 8

(f) Seed to lint ratio: 5:2

(g) Test weight:140 g

(h) Area: 10,000

Solution:

Yield of cotton Seed

$$= \frac{\begin{array}{c}\text{Area} \times \text{No. of sympodial branches/plant} \times \text{No. of bolls/branch} \\ \times \text{No. of seeds/locule} \times \text{Test weight}\end{array}}{\text{Spacing} \times 1000 \times 1000 \times 100 \times 10}$$

$$= \frac{10,000 \times 5 \times 4 \times 4 \times 8 \times 140}{0.9 \times 0.6 \times 1000 \times 1000 \times 100 \times 10} = \textbf{1.66 t/ha}$$

Yield of lint in t/ha $= \dfrac{\text{Yield of cotton seed in t/ha}}{\text{Seed to lint ratio}}$

$$= \frac{1.66 \times 2}{5} = \textbf{0.66 t/ha}$$

Yield of Seed Cotton = Yield of cotton seed + yield of lint

$$= 1.66 + 0.66 = \textbf{2.32 t/ha}$$

Ginning percentage $= \dfrac{\text{Yield of lint in t/ha} \times 100}{\text{Yield of Seed cotton in t/ha}}$

$$= \frac{0.66 \times 100}{2.32} = \textbf{28.45 \%}$$

Example 6 : Calculate the expected yield of maize grain in t/ha from the details mentioned below:

(a) Spacing: 75 cm × 30 cm

(b) Average number of cobs/plant: 2

(c) Average number of grain row /cob: 10

(d) Average number of seeds/grain row: 25

(e) Test weight of seeds:250 g

Solution:

$$\text{Yield} = \frac{\begin{array}{c}10{,}000 \times \text{No. of cobs/plant} \times \text{No. of grain row/cob} \\ \times \text{ No. of seeds/grain row} \times \text{Test weight}\end{array}}{\text{Spacing} \times 1000 \times 1000 \times 100 \times 10}$$

$$= \frac{10{,}000 \times 2 \times 10 \times 25 \times 250}{0.75 \times 0.30 \times 1000 \times 1000 \times 100 \times 10} = \textbf{5.56 t/ha}$$

Example 7: What would be the yield of rice grains in t/ha,if

(a) Number of panicles/m^2 :260

(b) Number of filled grains/panicle: 136

(c) Test weight: 20 g

Solution :

Yield

$$= \frac{10{,}000 \times \text{Number of panicles/m}^2 \times \text{Number of filled grains/panicle} \times \text{Test Weight}}{1000 \times 1000 \times 100 \times 10}$$

$$= \frac{10{,}000 \times 260 \times 136 \times 20}{1000 \times 1000 \times 100 \times 10} = \textbf{7.08 t/ha}$$

Example 8: From the following details, work out the grain yield of husked rice in t/ha, percentage of unfilled grains, fertility ratio and hulling percentage of rice

(a) Spacing: 20 cm × 20 cm

(b) Average number of effective tillers/hill: 9

(c) Average number of grains/panicle: 160

(d) Average number of unfilled grains/panicle:20

(e) Test weight : 22 g

(f) Yield of unhusked rice : 4.5 t/ha

Solution

Number of filled grains/panicle= Number of total grains/panicle-Number of unfilled grains/panicle

Number of filled grains/panicle = 120-60 = **140**

$$\textbf{Yield} = \frac{\text{Area x Number of effective tillers/hill} \times \text{Number of filled grains/panicle} \times \text{Test weight}}{0.2 \times 0.2 \times 1000 \times 1000 \times 100 \times 10}$$

$$\text{Yield} = \frac{10,000 \times 9 \times 140 \times 22}{0.2 \times 0.2 \times 1000 \times 1000 \times 100 \times 10} = \textbf{6.93 t/ha}$$

$$\text{Hulling Percentage} = \frac{\text{Yield of unhusked rice in t/ha} \times 100}{\text{Yield of husked rice in t/ha}}$$

$$= \frac{4.5 \times 100}{6.93} = \textbf{65.0\%}$$

$$\text{Fertility ratio} = \frac{\text{No. of filled grains/panicle}}{\text{No. of unfilled grains/panicle}}$$

$$= \frac{140}{20} = \textbf{7}$$

$$\text{Percentage of unfilled grains} = \frac{\text{No. of filled grains/panicle}}{\text{No. of total grains/panicle}} \times 100$$

$$= \frac{20 \times 100}{160} = \textbf{12.5\%}$$

Example 9: Estimate the yield of wheat grains and straw in t/ha from the observations given below:

(a) Spacing : 20 cm × 3 cm

(b) Average number of effective tillers/plant : 3

(c) Average number of grains/earhead: 32

(d) Test weight: 40 g

(e) Grain /Straw ratio: 1:1.5

Solution:

$$\textbf{Yield of Wheat grains} = \frac{\begin{array}{c}\text{Area} \times \text{Number of effective tillers/plant} \\ \times \text{ Number of grains/earhead} \times \text{ Test Weight}\end{array}}{\text{Spacing} \times 1000 \times 1000 \times 100 \times 10}$$

$$= \frac{10,000 \times 3 \times 32 \times 40}{0.20 \times 0.03 \times 1000 \times 1000 \times 100 \times 10} = \textbf{6.40 t/ha}$$

Yield of wheat straw $= \dfrac{\text{Grain Yield} \times 1}{\text{Grain /Straw ratio}}$

$$= \frac{6.40 \times 1}{1/1.5} = \textbf{9.6 t/ha}$$

Example 10: Find out the yield of jute fibre in q/ha from the following details:

(a) Spacing: 25 cm × 10 cm

(b) Average weight of matured plant:65 g

(c) Extractable fibre percentage: 7.5

Solution:

$$\text{Yield} = \frac{10,000 \times \text{Weight of matured plant} \times \text{Extractable fibre \%}}{\text{Spacing} \times 1000 \times 100 \times 10}$$

$$= \frac{10,000 \times 65 \times 75}{0.25 \times 0.1 \times 1000 \times 100 \times 10} = \textbf{19.5 q/ha}$$

Example 11: What would be the yield of jute in q/ha from the details are given below

(a) Spacing: 25 cm × 10 cm

(b) Average number of seeds/plant:185

(c) Test weight of seed: 5g

Solution:

$$\text{Yield} = \frac{\text{Area} \times \text{Number of seeds/plant} \times \text{Test weight of seed}}{\text{Spacing} \times 1000 \times 1000 \times 100}$$

$$= \frac{10,000 \times 185 \times 5}{0.25 \times 1000 \times 1000 \times 100} = \textbf{3.7 q/ha}$$

Example 12: Estimate the yield of groundnut pods, kernels and oil from the following particulars:

(a) Spacing : 50 cm × 25 cm

(b) Average number of matured pods/plant: 26

(c) Number of kernels/pod: 3

(d) Test weight of kernels: 650 g

(e) Shelling percentage: 60

(f) Percentage of oil in kernels: 45

Solution:

$$\text{Yield} = \frac{10,000 \times \text{No. of pods/plant} \times \text{No. of kernels/pod} \times \text{Test weight}}{\text{Spacing} \times 1000 \times 1000 \times 100}$$

$$= \frac{10,000 \times 26 \times 3 \times 650}{0.5 \times 0.25 \times 1000 \times 1000 \times 100} = \textbf{40.6 q/ha.}$$

$$\text{Yield of unshelled pods} = \frac{\text{Yield of kernels} \times 100}{\text{Shelling percentage}}$$

$$= \frac{40.6 \times 100}{60} = \textbf{67.6 q/ha}$$

Example 13: Find out the seed yield and oil yield of mustard from the following details:

(a) Spacing: 30 cm × 10 cm

(b) Number of branches/plant: 5

(c) Number of siliqua/branch: 20

(d) Number of seeds/siliqua: 15

(e) Test weight: 3g

(f) Area: 10 ha.

(g) Oil content in seed: 35%

Solution:

$$\text{Yield} = \frac{\text{Area} \times \text{No. of branches/plant} \times \text{No. of siliqua/branch} \times \text{No. of seeds/siliqua} \times \text{Test weight}}{\text{Spacing} \times 1000 \times 1000}$$

$$= \frac{10{,}000 \times 3 \times 5 \times 20 \times 15}{0.30 \times 0.10 \times 1000 \times 1000} = \textbf{15,000 kg/ha}$$

Mustard oil yield $= 15{,}000 \times \dfrac{35}{100} = \textbf{5250 kg}$

Example 14: Calculate the expected yield of maize grain from the following details:

(a) Spacing: 75 cm × 30 cm

(b) Number of cobs/plant:3

(c) Number of grain row/cob:10

(d) Number of grains/grain row: 20

(e) Test weight:220g

(f) Area: 5000 m²

Solution:

$$\text{Yield} = \frac{\text{Area} \times \text{No. of cobs/plant} \times \text{No. of grain row/cob} \times \text{No. of grains/grain row} \times \text{Test weight}}{\text{Spacing} \times 1000 \times 1000}$$

$$= \frac{5000 \times 3 \times 10 \times 20 \times 220}{0.75 \times 0.30 \times 1000 \times 1000} = \textbf{2933.3 kg}$$

Example 15: Calculate the expected yield of husked and unhusked rice from the following details:

(a) Spacing: 20 cm × 20 cm

(b) Number of panicles/hill: 10

(c) Number of total grains/panicle:200

(d) Percent filled grain:90

(e) Test weight: 17 g

(f) Spacing: 20 cm × 20 cm

(g) Hulling percentage:70

Solution:

$$\text{Total filled grains/panicle} = 200 \times \frac{90}{100} = 180$$

$$\text{\textbf{Yield}} = \frac{\text{Area} \times \text{Number of panicles/hill} \times \text{Number of filled grains/panicle} \times \text{Test weight}}{\text{Spacing} \times 1000 \times 1000}$$

$$= \frac{10{,}000 \times 10 \times 180 \times 17}{0.20 \times 0.20 \times 1000 \times 1000} = \textbf{7650 kg/ha}$$

Unhusked rice yield = Husked rice yield × Hulling percentage

$$= 7650 \times \frac{70}{100} = \textbf{5355 kg/ha}$$

Example 15: Estimate the tuber yield of potato from 1000m^2 area from the following observation:

(a) Average tuber weight: 40 g

(b) Number of tubers/plant:10

(c) Spacing: 50 cm × 30 cm

(d) Damage during harvest: 5%

Solution:

Yield =

$$\frac{\text{Area} \times \text{Tuber weight} \times \text{Undamaged tuber (\%)} \times \text{Number of tubers/plant}}{\text{Spacing} \times 1000 \times 1000 \times 100}$$

$$= \frac{1000 \times 40 \times 95 \times 10}{0.50 \times 0.30 \times 1000 \times 1000 \times 100} = \textbf{38 q}$$

⋮ EXERCISES ▶

1. Calculate the rice grain yield/m², if grain weight/panicle and number of panicles/m² is 3.5 g and 150, respectively.

2. Calculate the rice grain yield/ha, if grain weight/panicle and number of panicles/m² is 3.5 g and 200, respectively.

3. Calculate the grain yield of rice from the details mentioned below:
   (a) Average filled grains/panicle:160
   (b) Number of panicles/hill:9
   (c) Test weight:20 g
   (d) Spacing:20 cm × 15 cm

4. Calculate the expected yield of husk and unhusk rice from the following details:
   (a) Spacing: 20 cm × 20 cm
   (b) Number of panicles/hill:8
   (c) Total number of filled grains/panicle:300
   (d) Percent filled grain:85
   (e) Hulling percentage:65

5. Work out the grain yield of rice/ha. if
   (a) Average earhead density/m²:200
   (b) Average filled grains/panicle:220
   (c) Test weight:15 g

6. Calculate the expected yield of greengram in kg for 1000 m² area from the following observation:
   (a) Spacing: 30 cm × 10 cm
   (b) Number of pods/plant: 22
   (c) Number of seeds/pod: 11
   (d) Test weight: 31 g

7. Estimate the productivity of potato (q/ha) from the following observation:

   (a) Average tuber weight: 35 g

   (b) Number of tubers/plant: 11

   (c) Spacing: 50 cm × 20 cm

8. Estimate the tuber yield of potato from $1000m^2$ from the following observation:

   (a) Average tuber weight:30 g

   (b) Number of tubers/plant:11

   (c) Damage during harvest:6 %

9. Calculate the expected seed cotton and lint yield of cotton from the following details:

   (a) Spacing: 60 cm × 22.5 cm

   (b) Average sympodial branches/plant: 8

   (c) Average number of bolls/sympodia: 2

   (d) Boll weight: 3.9 g

   (e) Ginning percentage:33

   (f) Area: 1.5 ha.

10. Calculate the expected yield of mustard seed and oil yield of mustard from the following details:

    (a) Spacing: 30 cm × 10 cm

    (b) Number of branches/plant: 6

    (c) Number of siliqua/branch:18

    (d) Number of seeds/siliqua:14

    (e) Test weight: 3.1 g

    (f) Oil content in seed=35%

    (g) Area: 5 ha.

11. Calculate the expected yield of maize grain from the following details:
    (a) Spacing: 75 cm × 30 cm
    (b) Number of cobs/plant: 2
    (c) Number of grains/cob: 250
    (d) Test weight: 230 g
    (f) Area: 1 acre

12. Calculate the expected yield of maize grain from the following details:
    (a) Spacing: 75 cm × 30 cm
    (b) Number of cobs/plant: 2
    (c) Number of grain row/cob:10
    (c) Number of grains/grain row: 25
    (d) Number of grains/grain row:25
    (e) Test weight: 220 g
    (f) Area: 2.5 acre

13. Estimate the yield of wheat grain and straw in t/ha from the following details:
    (a) Row to row spacing: 18 cm
    (b) Number of effective tillers/hill: 80
    (c) Number of grains/panicle:30
    (d) Test weight: 40 g
    (e) Grain: Straw ratio: 1:1.5

14. Estimate the grain yield and unhusked rice from the following details:
    (a) Spacing: 20 cm × 20 cm
    (b) Number of effective tillers/hill:8
    (c) Number of grains/panicle:180
    (d) Number of unfilled grains/panicle:30
    (e) Test weight: 20 g
    (f) Hulling percentage: 65
    (g) Area: 1 ha.

15. Calculate the expected yield of rice in t/ha, if

    (a) Number of panicles/m^2: 250

    (b) Number of filled grains/panicle:140

    (c) Test weight: 16 g

16. Estimate the expected yield of cotton seed, lint, seed cotton and ginning percentage from the following details:

    (a) Area: 3 ha.

    (b) Spacing: 60 cm × 30 cm

    (c) Number of sympodialbranches/plant:3

    (d) Number of bolls/plant:3

    (e) Number of locules/boll:4

    (f) Number of seeds/locule:7

    (g) Seed to lint ratio:5:2

    (h) Test weight:110 g

17. Estimate the yield of sugarcane,cane juice, sugar and jaggari from the following details:

    (a) Average number of millable canes/clump:14

    (b) Average weight of canes: 850 g

    (c) Clump spacing: 90 cm × 60 cm

    (d) Extractable juice in cane: 70 %

    (e) Sugar recovery from juice: 16 %

    (f) Jaggari recovery from juice: 20%

    (g) Area: 10 ha.

⋮ ANSWERS ▶

1. 525 g   **2.** 700g   **3.** 9600 kg

4. Husked rice=7650 kg/ha; unhusked rice: 6502.50 kg/ha

5. 6600 kg   **6.** 250 kg   **7.** 385 kg

8. 31.02

9. Seed cotton yield= 4622.22 kg/ha; Lint yield= 1525.33 kg/ha

10. Seed yield: 7812 kg; Oil yield: 2968.6 kg

11. 2044.4 kg

12. 4888.9 kg

13. 5.33 t;8 t

14. 6 t; 3.9 t

15. 5.6 t

16. Cotton seed yield = 4.62 t; Lint yield =1.85 t; Seed cotton yield = 6.47 t

17. Sugarcasne yield = 2203.7t, juice yield = 1542.6t,sugar yield = 246.8; jaggery yield = 308.5t

FERTILIZER REQUIREMENT

Fertilizers are inorganic materials of a concentrated nature, applied to increase the supply of one or more of the essential nutrients. All the fertilizers can be grouped into following three groups:

1. **Straight Fertilizers:** Fertilizers which are used to supply only one primary plant nutrient are called straight fertilizers. For example, Urea (contains 46 % N), Single Super Phosphate contains 16 % P_2O_5), muriate of potash (contains 60 % K_2O), etc.

2. **Complex Fertilizers:** Complex fertilizers contain two or three primary plant nutrients of which two primary nutrients are in chemical combination. For example, diammonium phosphate (contain 18 % N and 46 % P_2O_5).

3. **Mixed fertilizers:** Mixed fertilizers have no definite chemical formula. These are simply a mechanical mixture or blends of different fertilizer materials.

On every fertilizer bag, 'grade' is printed. Fertilizer grade refers to the minimum percentage of N, P_2O_5 and K_2O contained in fertilizer material. For example, label on fertilizer bag 20-20-0 indicates that 100 kg fertilizer material contains 20 kg N, 20 kg P_2O_5 and no K_2O. Urea may be looked upon as 46-0-0 grade and DAP as 18-46-0.

The fertilizer ratio refers to the relative percentage of N, P_2O_5 and K_2O. A 6-24-24 grade has 1:4:4 ratio.

TABLE 1. Nutrient content of some fertilizers

Nature	Name of fertilizer	Nutrient fertilizer		
		N	P_2O_5	K_2O
Nitrogenous	Urea	46	-	-
	Calcium ammonium nitrate (CAN)	25	-	-
	Ammonium sulphate	20	-	-
	Ammonium chloride	25	-	-
	Ammonium nitrate	33.5	-	-
	Anhydrous ammonia	81.5	-	-
Phosphatic	Single super phosphate (SSP)	-	16	-
	Double super phosphate	-	32	-
	Triple super phosphate	-	48	-
	Rock phosphate	-	20	-
Potassic	Muriate of potash (MOP)	-	-	60
	Potassium sulphate	-	-	50
Complex	Diammonium phosphate (DAP)	18	46	-

Fertilizer calculations help to compute the correct amount of fertilizer material to be applied to a given area at the recommended rate. Fertilizer recommendations are given in kg of nutrient elements per hectare in order of N, P_2O_5 and K_2O. When the recommendation of fertilizer and fertilizer material are known, required quantities of fertilizer material for individual plot of the farm can be calculated.

For fertilizer calculations, following data are required:

1. Recommended rate of fertilizer
2. Nutrient content of fertilizer material
3. Area to be fertilized

5.1. FERTILIZER RQUIREMENT WITH STARIGHT FERTILIZER

Example 1: Calculate the amount of Urea, SSP and MOP to supply 40-40-40 kg N-P_2O_5-K_2O in 1 ha area.

Solution:

100 kg Urea contains	= 46 kg N
100 kg SSP contains	= 16 kg P_2O_5
and 100 kg MOP contains	= 60 kg K_2O

Requirement of urea to apply 1 kg N $= \dfrac{100}{46}$

$= 2.17$ kg

and Requirement of SSP to apply 1 kg P_2O_5 $= \dfrac{100}{16}$

$= 6.25$ kg

and Requirement of MOP to apply 1 kg K_2O $= \dfrac{100}{60}$

$= 1.67$ kg

Requirement of urea to apply 40 kg N = **2.17**[*] × 40 = 86.8 kg

Requirement of SSP to apply 40 P_2O_5 = **6.25**[*] × 40 = 250.0 kg

Requirement of MOP to apply 40 K_2O = **1.67**[*] × 40 = 66.8 kg

[* Students are suggested to remember these multiplying factors to speed up fertilizer calculations]

Example 2. Calculate the amount of Urea, SSP and MOP to supply N-P_2O_5-K_2O @ 40-40-40 kg per hectare in 5000 m^2 area.

Solution:

Requirement of urea to apply 40 kg N = 2.17 × 40 = 86.8 kg

Requirement of SSP to apply 40 P_2O_5 = 6.25 × 40 = 250.0 kg

Requirement of MOP to apply 40 K_2O = 1.67 × 40 = 66.8 kg

$$\text{Again, } 5000 \text{ m}^2 = \frac{5,000}{10,000}$$

$$= 0.5 \text{ ha}$$

For 5000 m², *i.e.* 0.5 ha,

Requirement of urea	$= 86.80 \times 0.5 \text{ kg} = 43.4 \text{ kg}$
Requirement of SSP	$= 250.00 \times 0.5 \text{ kg} = 125.0 \text{ kg}$
Requirement of MOP	$= 66.80 \times 0.5 \text{ kg} = 33.4 \text{ kg}$

Requirement of fertilizer material (kg)	= Recommended rate (kg/ha) × Multiplying factor × Area to be fertilised (ha)

Example 3. Calculate the amount of Urea, SSP and MOP to supply N-P_2O_5-K_2O @ 60-20-40 kg per hectare in 3 ha rice area.

Solution:

Requirement of urea = $2.17 \times 60 \times 3 = 390.6$ kg

Requirement of SSP = $6.25 \times 20 \times 3 = 375.0$ kg

Requirement of MOP = $1.67 \times 40 \times 3 = 200.4$ kg

Example 4. Calculate the requirement of ammonium sulphate to supply N @ 50 kg/ha for an area of 2000 m².

Solution:

Requirement of ammonium sulphate to supply 1 kg N = 100/20 = 5.0 kg

For 2000 m², *i.e.* 0.2 ha, requirement of ammonium sulphate

$$= 50 \times 5 \times 0.2 = 50.0 \text{ kg}$$

Example 5. The recommended dose of N for a wheat crop is 80 kg/ha. Find out the fertilizer requirement from the following fertilizer separately for a 2 ha area:

(*a*) Urea

(*b*) CAN

(*c*) Ammonium sulphate

(*d*) Ammonium chloride

(*e*) Ammonium nitrate

Solution:

Required amount of Urea, CAN, Ammonium sulphate, Ammonium chloride and Ammonium nitrate to get 1 kg of N (*i.e.* **multiplying factor**) = 2.17, 4, 5, 4 and 2.99

Requirement of urea = 80 × 2.17 × 2 kg = **347.2 kg**

Requirement of CAN = 80 × 4 × 2 kg = 640 kg

Requirement of Ammonium Sulphate = 80 × 5 × 2 kg = 800 kg

Requirement of Ammonium Chloride = 80 × 4 × 2 kg = 640 kg

Requirement of Ammonium Nitrate = 80 × 2.99 × 2 kg = 478.4 kg

Example 6. Calculate the fertilizer requirement (basal as well as top dress) for an area of 2.5 ha of rice in the form of Urea, SSP and MOP when N-P_2O_5-K_2O are applied @ 60-20-40 kg/ha. Assume that half of the urea and entire quantity of SSP and MOP are applied as basal and remaining half of the urea is top dressed at equal amount at maximum tillering and PI stage.

Solution:

Given that

N-P_2O_5-K_2O requirement for basal application = 30-20-40 kg/ha

N-P_2O_5-K_2O requirement for 1st top dressing = 15-0-0 kg/ha

N-P_2O_5-K_2O requirement for 2nd top dressing = 15-0-0 kg/ha

For basal application, requirement of

Urea = 30 × 2.17 × 2.5 kg		= 162.8 kg
SSP = 20 × 6.25 × 2.5 kg		= 312.5 kg
MOP = 40 × 1.67 × 2.5 kg		= 167.0 kg

For first top dressing requirement of urea = 15 × 2.17 × 2.5 kg

= 81.4 kg

For second top dressing requirement of urea $= 15 \times 2.17 \times 2.5$ kg

$\qquad\qquad\qquad\qquad\qquad\qquad\qquad\qquad\quad = 81.4$ kg

Example 7. A farmer planned to purchase 400 kg of Ammonium Sulphate. In the market Ammonium Sulphate was not available. Instead, Urea was available. Find out how much Urea the farmer had to purchase in place of Ammonium Sulphate.

Solution:

400 kg of Ammonium Sulphate will yield $= (400 \times 20/100)$ kg $= 80$ kg N

Amount of Urea to be purchased $\qquad = 80 \times 2.17$ kg

$\qquad\qquad\qquad\qquad\qquad\qquad\qquad\quad = 173.6$ kg

Example 8. An area of 40 m $\times$ 30 m is to be fertilized with Urea, SSP and MOP with N-P_2O_5-K_2O @ 40-20-20 kg/ha. Find out the quantity of fertilizer separately.

Solution:

Area to be fertilised $\qquad = 40$ m $\times$ 30 m

$\qquad\qquad\qquad\qquad\quad = 1200$ m^2

$\qquad\qquad\qquad\qquad\quad = 0.12$ ha

Requirement of Urea $\qquad = 2.17 \times 40 \times 0.12 = 10.2$ kg

and Requirement of SSP $\quad = 6.25 \times 20 \times 0.12 = 15.0$ kg

and Requirement of MOP $= 1.67 \times 20 \times 0.12 = 4.0$ kg

5.2. FERTILIZER REQUIREMENT WITH COMPLEX FERTILIZER

Calculation of straight fertilizer is very simple. But students encounter problem with complex fertilizer. However, if the problem is studied carefully it is as easy as the calculation with straight fertilizer.

Let us calculate the **amount of Urea and DAP to supply 20-40-0 and 10-40-0 N-P_2O_5-K_2O in 1 ha area.** We have already come to know that urea contains 46 % N and DAP contain 18 % N and 46 %

P_2O_5. Now, the question is whether we should calculate N or P_2O_5 first with DAP. *The normal convention is to calculate the amount of DAP to provide the required amount of P_2O_5. The amount of N which can not be met by DAP is replaced with urea or any other available nitrogenous fertilizer.* But what should we do *if calculated DAP provide higher amount of N than the required amount ?* Here are two options (*i*) Let DAP provide a bit higher amount of N (*ii*) Calculate the amount of N first and substitute the P_2O_5 which can not be met with DAP by SSP. Generally, people do not go for second option. However, for calculation point of view, both the issues will be discussed.

Here, to provide 40 kg P_2O_5, requirement of DAP = 40 × 2.17 kg = 86.8 kg
Amount of N that will be provided by this DAP = 86.8 × 18/100 kg

$$= 15.6 \text{ kg}$$

Thus, in the first case (20-40-0), DAP has provided 20 − 15.6 = 4.4 kg N less than the desired amount of N which should be substituted by Urea, *i.e.* the requirement of urea in the first case will be 9.5 kg (4.4 × 2.17 kg). However, in the second case (10-40-0), it will provide extra 15.6 − 10 = 5.6 kg N. Should we provide that extra N. I definitely will not provide that extra N and go for the second option.

What is the second option?

First, calculate the amount of DAP to provide 10 kg N. It will be 10 × 100/18 kg = 55.6 kg DAP.

Now, the amount of N that will be provided by this DAP = 55.6 × 46/100 kg

$$= 25.6 \text{ kg}$$

Thus, DAP has given 40 - 25.6 = 14.4 kg P_2O_5 less than the desired amount which I shall substitute by SSP or source of P_2O_5 available in the market. If it is substituted by SSP, the requirement will be = 14.4 × 6.25 kg = 90 kg

Thus, application of 25.6 kg DAP and 90 kg SSP will provide 10-40-0 kg N-P_2O_5-K_2O precisely in 1 ha area.

How to judge the nutrient to be substituted

(1) Calculate the ratio of nutrient content of the fertilizer to the amount of nutrients to be supplied. In the first case (20-40-0), nutrient content of DAP and fertilizer application rate are 18-46-0 and 20-40-0, respectively. Here ration of N is $18/20 = 0.90$ and ratio of P_2O_5 is $46/40 = 1.15$. In the second case (10-040-0), ratios N and P_2O_5, and 1.8 and 1.15, respectively.

(2) Substitute the nutrient first which has the highest value. In the first case (20-40-0), ratios for N and P_2O_5 are 0.90 and 1.15, respectively. Hence, calculation should be done for P_2O_5 first to meet the deficit N by urea. Similarly, for the second case (10-40-0), ratios for N and P_2O_5 are 1.80 and 1.15. So, calculation should be done for N to meet the deficit P_2O_5 by SSP.

Example 9. Calculate the requirement of Urea, DAP and MOP for 1 acre greengram crop, if the crop is fertilised with 13.7-35-10 kg N-P_2O_5-K_2O/ha.

Solution:

Here ration of N is $18/13.7 = 1.31$ and ratio of is $46/35 = 1.31$

Thus, calculation can be made either for N or for P_2O_5

For 1 ha area:

Let us calculate N first,

Here, requirement of DAP to give 13.7 kg N/ha = $13.7 \times 100/18 = 76.1$ kg which will give $76.1 \times 46/100 = 35$ kg P_2O_5. Thus no requirement of additional phosphatic source.

Let us calculate P_2O_5 first,

Here, requirement of DAP to give 5 kg P_2O_5/ha = $35 \times 100/46 = 76.1$ kg which will give $76.1 \times 18/100 = 13.7$ kg N. Thus no requirement of additional nitrogenous fertilizer.

Requirement of MOP = 10×1.67 kg = 16.7 kg

For 1 acre area :

  Requirement of Urea = Not required

  DAP = 76.1/2.5 = 30.4 kg

  MOP = 16.7/2.5 = 6.7 kg

Example 10. Calculate the amount of Urea, DAP and MOP to supply 40-40-40 kg N-P_2O_5-K_2O in 1 ha area.

Solution:

Requirement of DAP = 40 × 2.17 kg

= 86.80 kg

Amount of N supplied by 86.80 kg N= 86.8 × 18/100 kg

= 15.6 kg

Requirement of Urea = (40 − 15.6) × 2.17 kg

= 24.4 × 2.17 kg

= 52.9 kg

Requirement of MOP = 40 × 1.67 kg

= 66.8 kg

Example 11. Calculate the amount of Urea, DAP and MOP to supply N-P_2O_5-K_2O @ 40-40-40 kg per hectare in 5,000 m^2 area.

Solution:

Requirement of DAP/ha = 40 × 2.17 kg

= 86.80 kg

Amount of N supplied by 86.80 kg N= 86.8 × 18/100 kg

= 15.6 kg

Requirement of Urea/ha = (40 − 15.6) × 2.17 kg

= 24.4 × 2.17 kg

= 52.9 kg

Requirement of MOP/ha = 40 × 1.67 kg

= 66.8 kg

Hence, for 5,000 m², *i.e.* 0.5 ha,

Requirement of Urea	$= 52.9 \times 0.5$ kg $= 26.45$ kg
Requirement of SSP	$= 86.8 \times 0.5$ kg $= 43.4$ kg
Requirement of MOP	$= 66.8 \times 0.5$ kg $= 33.4$ kg

Example 12. Calculate the amount of Urea, DAP and MOP to supply N-P_2O_5-K_2O @ 60-20-40 kg per hectare in 3 ha rice area.

Solution:

Requirement of DAP/ha	$= 20 \times 2.17$ kg
	$= 43.4$ kg
Amount of N supplied by 43.4 kg N	$= 43.4 \times 18/100$ kg
	$= 7.8$ kg
Requirement of Urea/ha	$= (60 - 7.8) \times 2.17$ kg
	$= 113.3$ kg
Requirement of MOP/ha	$= 40 \times 1.67$ kg
	$= 66.8$ kg

Hence, for 3 ha,

Requirement of Urea	$= 113.3 \times 3$ kg $= 339.9$ kg
Requirement of SSP	$= 43.4 \times 3$ kg $= 130.2$ kg
Requirement of MOP	$= 66.8 \times 3$ kg $= 200.4$ kg

Example 13. Calculate the fertilizer requirement (basal as well as top dress) for an area of 2.5 ha of rice in the form of Urea, DAP and MOP when N-P_2O_5-K_2O are applied @ 60-20-40 kg/ha. Assume that half of the urea and entire quantity of DAP and MOP are applied as basal and remaining half of the urea is top dressed at equal amount at maximum tillering and PI stage.

Solution:

Given that

N-P_2O_5-K_2O requirement for basal application = 30-20-40 kg/ha

N-P_2O_5-K_2O requirement for 1ˢᵗ top dressing = 15-0-0 kg/ha

N-P_2O_5-K_2O requirement for 2nd top dressing = 15-0-0 kg/ha

Hence, for basal application,

Requirement of DAP/ha	= 20 × 2.17 kg = 43.4 kg
43.4 kg DAP will give	= (43.4 × 18/100) kg N
	= 7.8 kg N
Requirement of Urea/ha	= (30 − 7.8) × 2.17 kg
	= 48.2 kg
Requirement of MOP/ha	= 40 × 1.67 kg
	= 66.8 kg

For 2.5 ha area, for basal application, requirement of -

Urea = 48.2 × 2.5 kg = 120.5 kg

DAP = 43.4 × 2.5 kg = 108.5 kg

MOP = 66.8 × 2.5 kg = 167.0 kg

and for first top dressing requirement of urea

= 15 × 2.17 × 2.5 kg = 81.4 kg

and for second top dressing requirement of urea

= 15 × 2.17 × 2.5 kg = 81.4 kg

Example 14. A farmer planned to purchase 400 kg of SSP. In the market SSP was not available. Instead, DAP was available. Find out how much DAP the farmer had to purchase in place of SSP. Also, calculate the amount of urea he saves due to purchase of DAP.

Solution:

400 kg of SSP will yield	= (400 × 16/100) kg = 64 kg P_2O_5
Amount of DAP to be purchased	= 64 × 2.17 kg
	= 138.9 kg
138.9 kg of DAP will yield	= (138.9 × 18/100) kg = 25 kg N
	= 25 × 2.17 kg Urea
	= 54.25 kg Urea

As such, the farmer will save 54.25 kg urea

Example 15. An area of 40 m × 30 m is to be fertilized with Urea, DAP and MOP with N-P$_2$O$_5$-K$_2$O @ 40-20-20 kg/ha. Find out the quantity of fertilizer separately.

Solution:

Area to be fertilised	= 40 m × 30 m
	= 1200 m^2
	= 0.12 ha

Requirements/ha

DAP	= 2.17 × 20 kg = 43.4 kg

[43.4 kg DAP will yield = 43.4×18/100 = 7.8 kg N]

Urea	= (40 – 7.8) × 2.17 kg = 69.9 kg
MOP	= 1.67 × 20 kg = 33.4 kg

Requirement for 0.12 ha

Urea	= 69.9 × 0.12 kg = 8.4 kg
DAP	= 43.3 × 0.12 kg = 5.2 kg
MOP	= 33.4 × 0.12 kg = 4.0 kg\

5.3. FERTILIZER REQUIREMENT WITH MIXED FERTILIZER

Mixed fertilizers are physical mixtures of fertilizer materials containing two or more major plant nutrients. Mixed fertilizer materials are prepared by thoroughly mixing ingredients either mechanically known as 'factory made' or manually, known as 'farm mixed'. Every fertilizer mixture is solid with a declared fertilizer grade.

Different materials are needed for production of mixed fertilizer *viz.*, supplies of plant nutrients, conditioners, neutralisers of residual acidity and filler. Plant nutrients are supplied by fertilizers. Conditioners are added to fertilizer mixture during their preparation for reducing hygroscopicity and improving physical condition. Such materials may be tobacco stem, peat, groundnut hull and paddy husk etc. If fertilizers used are acidic (like ammonium sulphate, urea) in nature, a basic material like dolomitic lime stone is added as neutraliser to countract the acidity. A filler is the make weight material added to a fertilizer mixture. The common filler material used are sand, soil, ground coal ash and various other waste products.

Example 16. Find out the fertilizer materials needed for preparation of 1000 kg of mixed fertilizer of 8-8-8 using Urea, SSP and MOP.

Solution:

To produce 100 kg mixture,

Requirement of Urea	$= 8 \times 2.17$ kg $= 17.36$ kg
Requirement of SSP	$= 8 \times 6.25$ kg $= 50.0$ kg
Requirement of MOP	$= 8 \times 1.67$ kg $= 13.36$ kg
Filler materials	$= 100 - (17.36 - 50.00 - 13.36)$
	$= 19.28$ kg

To produce 1000 kg mixture, requirement of

Urea	$= 17.36 \times 1000/100$ kg $= 173.6$ kg
SSP	$= 50.00 \times 1000/100$ kg $= 500.0$ kg
MOP	$= 13.36 \times 1000/100$ kg $= 133.6$ kg
Filler	$= 19.28 \times 1000/100$ kg $= 192.8$ kg

Example 17. Find out the fertilizer materials needed for preparation of 1000 kg of field mixed fertilizer of 8-8-8 using Urea, DAP and MOP.

Solution:

To produce 100 kg mixture,

Requirement of DAP	$= 8 \times 2.17$ kg $= 17.36$ kg
Requirement of Urea	$= (8 - 17.36 \times 18/100) \times 2.17$ kg $= 10.60$ kg
Requirement of MOP	$= 8 \times 1.67$ kg $= 13.36$ kg
Filler materials	$= 100 - (17.36 - 10.60 - 13.36)$
	$= 58.68$ kg

To produce 1000 kg mixture, requirement of

Urea	$= 10.60 \times 1000/100$ kg	$= 106.0$ kg
SSP	$= 17.36 \times 1000/100$ kg	$= 173.6$ kg
MOP	$= 13.36 \times 1000/100$ kg	$= 133.6$ kg
Filler	$= 58.68 \times 1000/100$ kg	$= 586.8$ kg

Example 18. It is recommended that in *kharif* rice, farmers should apply N-P_2O_5-K_2O @ 60-20-40 kg/ha. Entire quantity of phosphatic and potassic fertilizers should be applied as basal and remaining half Urea should be top dressed at equal amount at maximum tillering and PI stage. Suggest a fertilizer mixture with Urea, SSP and MOP to apply fertilizer in 1 *bigha* (7.5 *bigha* = 1 ha) area. What is the advantage of this mixture?

Solution:

Basal requirement of N-P_2O_5-K_2O/ha $= 30$-20-40

Hence, requirement of N-P_2O_5-K_2O/bigha $= (30/7.5)$-$(20/7.5)$-$(40/7.5)$

$$= 4\text{-}2.67\text{-}5.33$$

So the suggested grade is 4-2.67-5.33 or multiple of it

If a grade of 4-2.67-5.33 is prepared,

Requirement of Urea	$= 4 \times 2.17 = 8.68$ kg
Requirement of SSP	$= 2.67 \times 6.25 = 16.69$ kg
Requirement of MOP	$= 5.33 \times 1.67 = 8.90$ kg
Total (without filler)	$= 34.27$ kg

Thus, another grade of 8 – 5.34 – 10.66 is possible with total fertilizer material of $34.27 \times 2 = 68.54$ kg

Let us prepare a mixture of grade 8 – 5.34 – 10.66

Now, Requirement of urea	$= 8 \times 2.17 = 17.36$ kg
Requirement of SSP	$= 4.34 \times 6.25 = 10.60$ kg
Requirement of MOP	$= 10.66 \times 1.67 = 17.80$ kg
Filler materials	$= 100 - (17.36\text{–}33.38 - 17.80)$
	$= 31.46$ kg

Thus 100 kg mixture of this grade will give 8 kg N, 5.34 kg P_2O_5 and 10.66 kg K_2O or a 50 kg mixture bag of this grade will give 4 kg N, 2.67 kg P2O5 and 5.33 kg K2O which can be easily applied to 1 *bigha kharif* rice as basal fertilizer.

By this time, the advantage of this mixture is clear to you. Since the farmers are not well verged with fertilizer calculation procedure, the fertilizer mixture specific to the major crops of the each region will help farmers for balanced and precise application of fertilizers.

Example 19. A farmer was taught to prepare a field mixture of grade 16-10.68-21.32 (which he needs to apply @ 25 kg/bigha) for basal application in his 40 *bigha kharif* rice crop with Urea, DAP and MOP. However, he wrongly prepared 500 kg field mixture of grade 15-15-15. How will he convert his mixture to 16-10.68-21.32 to apply in 40 *bigha* rice crop?

Solution:

Requirement of 16-10.68-21.32 grade mixture of the farmer

$$= 25 \times 40 \text{ kg} = 1000 \text{ kg}$$

The farmer needs 160 kg N, 106.8 kg P_2O_5 and 213.2 kg K_2O for 40 *bigha*

From the 500 kg mixture of 15-15-15 grade, he will get 75 kg N, 75 kg P_2O_5 and 75 kg K_2O.

Thus, remaining $(160 - 75) = 85$ kg N, $(106.8 - 75) = 31.8$ kg P_2O_5 and $(213.2 - 75) = 138.2$ kg K_2O must be supplied by the available stock of Urea, DAP and MOP.

Requirement of DAP = 31.8 × 2.17 kg	= 69.0 kg
Requirement of Urea = (85 − 69.0 × 0.18) × 2.17 kg	= 157.5 kg
Requirement of MOP = 138.2 × 1.67 kg	= 230.8 kg
Material from 10-10-10 grade	= 500.0 kg
Filler = (1000 − 500 − 230.8 − 157.3 − 69.0) kg	= 42.7 kg
Total	= 1000 kg

5.4. FERTILIZER RECOMMENDATION IN TERMS OF N, P AND K

Sometimes, Phosphorus and Potassium content in phosphatic and potassic fertilizers are expressed in terms of P and K, respectively. To convert P_2O_5 and K_2O to P and K, they should be divided by 2.29 and 1.2, respectively.

Example 20. The recommended rate of N, P_2O_5 and K_2O of upland direct seeded rice is 40-20-20 kg/ha. Express the recommendation in terms of N, P and K.

Solution:

$$20 \text{ kg } P_2O_5 \approx 20/2.29 \text{ kg P} \quad = 8.73 \text{ kg P}$$
$$20 \text{ kg } K_2O \approx 20/1.67 \text{ kg K} \quad = 11.98 \text{ kg K}$$

Recommended fertilizer rate is 40-8.73-11.98 N-P-K kg/ha

Example 21. Calculate the fertilizer requirement in the form of Urea, SSP and MOP for an area of 0.5 ha. The fertilizer rate is 40-10-10 N-P-K kg/ha.

Solution:

Recommended rate	$= 40\text{-}10\text{-}10$ N-P-K kg/ha
	$\approx 40\text{-}22.9\text{-}12.9$ N-P_2O_5-K_2O kg/ha
Requirement of urea	$= 40 \times 2.17 \quad = 86.8$ kg
SSP	$= 22.9 \times 6.25 = 143.1$ kg
MOP	$= 12.0 \times 1.67 = 20.0$ kg

Example 22. Calculate the fertilizer requirement in the form of Urea, DAP and MOP for an area of 0.5 ha. The fertilizer rate is 40-10-10 N-P-K kg/ha.

Solution:

Recommended rate	$= 40\text{-}10\text{-}10$ N-P-K kg/ha
	$\approx 40\text{-}22.9\text{-}12.9$ N-P_2O_5-K_2O kg/ha

Requirement of DAP $= 22.9 \times 2.17 = 49.7$ kg

Urea $= (40 - 49.7 \times 18/100) \times 2.17 = 67.4$ kg

MOP $= 12.0 \times 1.67 = 20.0$ kg

⋮EXERCISES ▶

1. Calculate the amount of Urea,SSP and MOP to supply 60:30:30 kg/ha $N:P_2O_5:K_2O$ in one hectare.

2. Calculate the amount of Urea,SSP and MOP to supply 60:30:30 kg/ha $N:P_2O_5:K_2O$ in 40 m × 30 m area.

3. Calculate the fertilizer requirement in the form of Urea,SSP and MOP for an area of 0.6 ha. The fertilizer rate is 40:20:20 $N:P_2O_5:K_2O$ kg/ha

4. A farmer planned to purchase 350 kg of CAN. In the market CAN was not available instead urea was available. Find out how much Urea the farmer had to purchase in place of CAN.

5. An area of 80 m × 50 m is to be fertilized with Urea,SSP and MOP with $N:P_2O_5:K_2O$ at the rate of 80:40:40 kg/ha. Find out the quantity of fertilizer separately.

6. A farmer had a stock of DAP. Instead of SSP, he wants to apply DAP with $N:P_2O_5:K_2O$ at the rate of 80:40:40 kg/ha. Find out the quantity of fertilizers for 4000m² area.

7. Calculate the amount of Urea, DAP and MOP to supply $N:P_2O_5:K_2O$ at the rate of 40:40:40 kg/ha. in 2 acre area.

8. Calculate the fertilizer requirement (basal as well as top dress) of rice crop for an area of 10 ha. in the form of Urea, SSP and MOP when $N:P_2O_5:K_2O$ are applied at the rate of 60:20:20 kg/ha. Assume that half of the urea and entire quantity of SSP and MOP are applied as basal and remaining half of the Urea is top dressed at equal amount at tillering and Panicle Initation stage.

9. Calculate the fertilizer requirement (basal as well as top dress) of rice crop for an area of 10 ha. in the form of Urea, DAP and MOP

when $N:P_2O_5:K_2O$ are applied at the rate of 60:20:20 kg/ha. Assume that half of the urea and entire quantity of DAP and MOP are applied as basal and remaining half of the Urea is top dressed at equal amount at tillering and Panicle Initation stage.

10. 100:50:50 $N:P_2O_5:K_2O$ kg/ha has to be applied on 1.5 ha. area of maize crop. Find out the quanity of Urea, DAP and Potassium Sulphate.

11. Calculate the fertilizer requirement for an area of 3ha. in the form of DAP,CAN and MOP when when $N:P_2O_5:K_2O$ are applied at the rate of 60:30:30 kg/ha.

12. Find out the cost of 1 kg N, P_2O_5 and K_2O if the cost of 1 kg urea,SSP and MOP are Rs.12,10 and 15,respectively.

⋮ ANSWERS ▶

1. Urea=130.4 kg; SSP=187.5 kg; MOP=50 kg
2. Urea=15.65 kg; SSP=22.50 kg; MOP=6 kg
3. Urea=52.1 kg; SSP=75 kg; MOP=20 kg
4. 189.9
5. Urea=69.44 kg; SSP=100 kg; MOP=26.72 kg
6. Urea=55.88 kg; SSP=34.72 kg; MOP=26.72 kg
7. Urea=42.32 kg; DAP=69.44 kg; MOP=53.44 kg
8. **Basal:**
   Urea=651 kg; SSP=1250 kg; MOP=668kg
   Top dressing : Urea at maximum tillering stage = 325.5 kg
   Urea at panicle initation stage = 325.5 kg
9. **Basal:**
   Urea=651 kg; SSP=1250 kg; MOP=668kg
   Top dressing : Urea at maximum tillering stage = 325.5 kg
   Urea at panicle initation stage = 325.5 kg

10. Urea=261.9 kg;   DAP=162.75 kg;   Potassium Sulphate=150 kg
11. CAN=597.4 kg;   DAP=195.3 kg;   MOP=150 kg
12. N= Rs. 26.1, P_2O_5 = Rs.143 and K_2O = 30.0

NUTRIENT UPTAKE AND NUTRIENT USE EFFICIENCY

Efficiency is defined as the amount of product (output) per unit input consumed. Usually, fertilizer increases the yield of crops. As such, farmer's willingness to use excess fertilizer has been increased. However, with the increase in fertilizer dose, rate of increase in crop yield decreases. Thus, with the increase in fertilizer dose, efficiency of applied nutrient is decreased. In this chapter, attempts are made to explain the concept of nutrient uptake and efficiencies of applied nutrients.

6.1. NUTRIENT UPTAKE

Nutrient uptake is the amount of nutrient taken up by the crop. It is computed as follows:

Nutrient Uptake (kg/ha)

$$= \frac{\text{Per cent nutrient content in grain or straw} \times \text{yield (kg/ha)}}{100}$$

Example 1. Grain and straw yield and nitrogen content of rice with 25 kg N/ha are given below. Calculate N uptake by the rice plant.

Nitrogen applied (kg/ha)	Grain yield (kg/ha)	Straw yield (kg/ha)	N content (%)	
			Grain	Straw
25	3,645	4,790	1.13	0.38

Solution :

Nitrogen uptake by grain (kg/ha) $= \dfrac{1.13 \times 3645}{100}$

$= 41.2$

Nitrogen uptake by straw (kg/ha) $= \dfrac{0.38 \times 4790}{100}$

$= 18.2$

Total N uptake $= (41.2 + 18.2)$ kg/ha

$= 59.4$ kg/ha

6.2. NUTRIENT USE EFFICIENCY

Nutrient Use Efficiency is judged by Agronomist, Soil Scientists and Physiologists differently with different indices.

6.2.1. Agronomic Efficiency (AE)

Agronomic Efficiency is the additional grain yield produced due to application of nutrients over unfertilized control per unit of nutrient applied. It is expressed on kg/g.

$$\text{Agronomic Efficiency (AE)} = \frac{GY_n - GY_0}{N_a}$$

Where,

GY_n = Grain or economic yield with nutrients

GY_0 = Grain or economic yield without nutrients

N_a = Nutrient applied

Example 2. Calculate the Agronomic Nitrogen Use Efficiency from the following information:

Nitrogen applied (kg/ha)	Grain yield (kg/ha)
0	3 490
25	3 645

Solution :

Given that,

$GY_n = 3645$ kg/ha

$GY_0 = 3490$ kg/ha

$N_a = 25$ kg/ha

Agronomic Efficiency (AE) $= \dfrac{GY_n - GY_0}{N_a}$

$= \dfrac{3645 - 3490}{25}$

$= \dfrac{155}{25}$

$= 6.2$ kg/kg

6.2.2. Physiological Efficiency (PE)

Physiological Efficiency is the additional biological yield produced due to application of nutrients over unfertilized control per unit of additional nutrient uptake over unfertilized control. It is expressed in kg/kg.

Physiological Efficiency (PE) $= \dfrac{BY_n - BY_0}{NU_n - NU_0}$

Where,

BY_n = Biological yield with nutrients

BY_0 = Biological yield without nutrients

NU_n = Nutrient uptake with nutrients

NU_0 = Nutrient uptake without nutrients

Example 3. Calculate the Physiological Nitrogen Use Efficiency from the following information:

Nitrogen applied (kg/ha)	Grain yield (kg/ha)	Straw yield (kg/ha)	N-uptake (kg/ha)	
			Grain	Straw
0	3,490	4,710	35.85	15.88
25	3,645	4,790	41.20	18.20

Solution :

Given that,

$$BY_n = (3,645 + 4,790) \text{ kg/ha}$$
$$= 8,435 \text{ kg/ha}$$

$$BY_0 = (3,490 + 4,710) \text{ kg/ha}$$
$$= 8,200 \text{ kg/ha}$$

$$NU_n = 41.20 + 18.20$$
$$= 59.40 \text{ kg/ha}$$

$$NU_0 = 35.85 + 15.88$$
$$= 51.73 \text{ kg/ha}$$

$$\text{Physiological Efficiency (PE)} = \frac{BY_n - BY_0}{NU_n - NU_0}$$

$$= \frac{8,435 - 8,200}{59.40 - 51.73}$$

$$= \frac{8\cancel{4}5\text{-}8\cancel{2}0}{5\cancel{9}0\text{-}5\cancel{1}7}$$

$$= 30.64 \text{ kg/kg}$$

6.2.3. Agro-physiological Efficiency (APE)

Agro-physiological Efficiency is the additional grain or economic yield produced due to application of nutrients over unfertilized control per unit of additional nutrient uptake over unfertilized control per unit of additional nutrient uptake over unfertilized control. It is expressed in kg/kg.

$$\text{Agro-physiological Efficiency (APE)} = \frac{GY_n - GY_0}{NU_n - NU_0}$$

Where,

GY_n = Grain yield with nutrients

GY_0 = Grain yield without nutrients

NU_n = Nutrient uptake with nutrients

NU_0 = Nutrient uptake without nutrients

Example 4. Calculate the Agro-physiological Nitrogen Use Efficiency from the following information:

Nitrogen applied (kg/ha)	Grain yield (kg/ha)	Straw yield (kg/ha)	N-uptake (kg/ha)	
			Grain	Straw
0	3,490	35.85	15.88	
25	3,645	41.20	18.20	

Solution:

Given that,

GY_n = 3,645 kg/ha

GY_0 = 3,490 kg/ha

NU_n = 41.20 + 18.20

= 59.40 kg/ha

NU_0 = 35.85 + 15.88

= 51.73 kg/ha

$$\text{Agro-physiological Nitrogen Use Efficiency} = \frac{GY_n - GY_0}{NU_n - NU_0}$$

$$= \frac{3,645 - 3,490}{59.40 - 51.73}$$

$$= \frac{155}{7.67}$$

$$= 20.21 \text{ kg/kg}$$

6.2.4. Apparent Recovery Efficiency (ARE)

Apparent Recovery Efficiency is the additional nutrient uptake over unfertilised control per unit of nutrient applied. It is expressed in percentage.

$$\text{Apparent Recovery Efficiency (ARE)} = \frac{BY_n - BY_0}{N_a} \times 100 \%$$

Example 5: Calculate the Apparent Recovery Efficiency of Nitrogen from the following information:

Nitrogen applied (kg/ha)	N-uptake (kg/ha)	
	Grain	Straw
0	35.85	15.88
25	41.20	18.20

Solution:

Given that,

NU_n = 41.20 + 18.20

= 59.40 kg/ha

NU_0 = 35.85 + 15.88

= 51.73 kg/ha

N_a = 25 kg/ha

$$\text{Apparent recovery efficiency of N} = \frac{NU_n - NU_0}{N_a} \times 100\%$$

$$= \frac{59.40 - 51.73}{25} \times 100\%$$

$$= \frac{7.67}{25} \times 100\%$$

$$= 30.68\%$$

6.2.5. Utilization Efficiency (UE)

Utilization Efficiency is the additional biological yield over unfertilized control per unit of nutrient applied. It is expressed in kg/kg.

$$\text{Utilization Efficiency (UE)} = \frac{BY_n - BY_0}{N_a}$$

$$= PE \times ARE$$

Example 6: Calculate the Utiliztion Efficiency of Nitrogen from the following information:

Nitrogen applied (kg/ha)	Yield (kg/ha)	
	Grain	Straw
0	3,490	4,710
25	3,645	4,790

Solution :

Given that,

BY_n = (3,645 + 4,790) kg/ha

= 8,435 kg/ha

BY_0 = (3,490 + 4,710) kg/ha

= 8,200 kg/ha

N_a = 25 kg/ha

$$\text{Utilization efficiency of N} = \frac{BY_n - BY_0}{N_a}$$

$$= \frac{8{,}435 - 8{,}200}{25}$$

$$= \frac{235}{25}$$

$$= 9.4 \text{ kg/kg}$$

6.2.6. Nutrient Efficiency Ratio (NER)

Nutrient Efficiency Ratio is the total biomass produced per unit of nutrient uptake. It is expressed in kg/kg.

$$\text{Nutrient Efficiency Ratio (NER)} = \frac{BY}{N}$$

Where,

BY = Biological Yield

NU = Nutrient Uptake

Example 7. Calculate the Nutrient Efficiency Ratio from the following information:

Nitrogen applied (kg/ha)	Grain yield (kg/ha)	Straw yield (kg/ha)	N-uptake (kg/ha)	
			Grain	Straw
25	3,645	4,790	41.20	18.20

Solution:

Given that,

BY	= (3,645 + 4,790) kg/ha
	= 8,435 kg/ha
NU	= 41.20 + 18.20
	= 59.40 kg/ha

Nitrogen Efficiency ratio $= \dfrac{BY}{NU}$

$$= \dfrac{8,435}{59.40}$$

$$= 142 \text{ kg/kg}$$

6.2.7. Nutrient Increment Efficiency (NIE)

Nutrient Increment Efficiency (NIE) is the additional grain or economic yield over the previous level of nutrient per unit of proceeding level of grain. It is expressed in kg/kg.

Nutrient Increment Efficiency (NIE) $= \dfrac{Y_n - Y_{n-1}}{Y_{n-1}}$

Where,

GY_n = Grain yield or economic yield with N_n amount of nutrient

GY_{n-1} = Grain yield or economic yield with N_{n-1} amount of nutrient

Example 8. Calculate the Nutrient Increment Efficiency of Nitrogen from the following information:

Nitrogen applied (kg/ha)	Grain yield (kg/ha)	Grain uptake (kg/ha)
25	3,645	41.2
50	4,025	47.4

Solution:

Given that,

Y_n = 4025 kg/ha

Y_{n-1} = 3645 kg/ha

N_n = 50 kg/ha

N_{n-1} = 25 kg/ha

$$\text{Nutrient Increment Efficiency of N} = \frac{Y_n - Y_{n-1}}{Y_{n-1}}$$

$$= \frac{4,025 - 3,645}{3,645}$$

$$= \frac{380}{3,645}$$

$$= 0.10 \text{ kg/kg}$$

6.2.8. Partial Production Efficiency (PPE)

Partial Production Efficiency is the additional grain or economic yield over the previous level of nutrient per unit of additional nutrient applied over preceding level of nutrient. It is expressed in kg/kg.

$$\text{Partial Production Efficiency (PPE)} = \frac{GY_n - GY_{n-1}}{N_n - N_{n-1}}$$

Where,

GY_n = Grain yield or economic yield with N_n amount of nutrient

GY_{n-1} = Grain yield or economic yield with N_{n-1} amount of nutrient

Example 9. Calculate the Partial Production Efficiency of Nitrogen from the following information:

Nitrogen applied (kg/ha)	Grain yield (kg/ha)	Grain uptake (kg/ha)
25	3,645	41.2
50	4,025	47.4

Solution:

Given that,

Y_n = 4025 kg/ha

Y_{n-1} = 3645 kg/ha

N_n = 50 kg/ha

N_{n-1} = 25 kg/ha

$$\text{Partial Production Efficiency of N} = \frac{GY_n - GY_{n-1}}{N_n - N_{n-1}}$$

$$= 15.2 \text{ kg/kg}$$

6.2.9. Economic Nutrient Use Efficiency (ENUE)

Economic Nutrient Use Efficiency (ENUE)

$$= \frac{\text{Grain or economic yield}}{\text{Amount invested on the nutrient}}$$

Example 10. Calculate the Economic Nitrogen Use Efficiency from the following information:

Nitrogen applied (kg/ha)	Grain yield (kg/ha)	Cost of Urea (¹ /kg)
25	3,645	12

Solution :

Cost of 1 kg urea	= ¹ 12.00
Cost of 1 kg N	= ¹ 12 × 2.17
	= ¹ 26.04
+. Cost of 25 kg N	= ¹ 26.04 × 25
	= ¹ 651.00

+. Economic Nutrient Use Efficiency (ENUE) = $\dfrac{3,645}{651}$

= 5.60 kg grain/ ¹ invested in Nitrogen

6.2.10. Nutrient harvest index (NHI)

Nutrient Harvest Index is the ratio of nutrient uptake by grain (or economic yield) to total nutrient uptake. It can also be expressed in percentage.

Nutrient Harvest Index

$$= \frac{\text{Nutrient uptake by grain (or economic yield)}}{\text{Total nutrient uptake for production of biological yield}}$$

Example 11. Calculate the Nitrogen Harvest Index at 0 and 25 kg/ha from the following information:

Nitrogen applied (kg/ha)	N – uptake (kg/ha)	
	Grain	Straw
0	35.85	15.88
25	41.20	18.20

Solution:

$$\text{Nutrient Harvest Index} = \frac{\text{N uptake by grain}}{\text{Total N uptake by plant}}$$

$$= \frac{35.85}{35.85 + 15.88}$$

$$= 0.6930$$

$$= 69.30 \%$$

6.2.11. Partial Nutrient Balance (PNB) or Nutrient Removal Ratio

Partial Nutrient Balance is the ratio of nutrient uptake by grain to nutrient applied. It is expressed in kg/kg.

$$\text{Partial Nutrient Balance} = \frac{\text{Nutrient uptake by grain (or economic yield)}}{\text{Total nutrient applied}}$$

Example 12. Calculate the Nitrogen Removal Ratio from the following information:

Nitrogen applied (kg/ha)	N uptake by grain (kg/ha)
25	41.2

Solution:

$$\text{Nitrogen Removal ratio} = \frac{41.2}{25}$$

$$= 1.65 \text{ kg/kg}$$

⋮ EXERCISES ▶

1. Seed and straw yield and nutrient content of rice with 0 and 30 kg N/ha are given below:

Nitrogen applied (kg/ha)	Seed yield (q/ha)	Stover yield (q/ha)	N-content (%)	
			Grain	Stover
0	7.45	12.80	1.93	0.68
30	9.09	15.53	2.19	0.77

Calculate

(a) N uptake by rice at 0 and 30 kg N/ha

(b) Agronomic Nitrogen Use Efficiency

(c) Physiological Nitrogen Use Efficiency

(d) Agro-physiological Nitrogen Use Efficiency

(e) Apparent Recovery Efficiency of Nitrogen

(f) Nitrogen Harvest Index

(g) Utilization Efficiency

(h) Nitrogen Efficiency Ratio at 30 kg N/ha

(i) Partial Nitrogen Balance

2. Calculate the Economic Nutrient Use Efficiency of Q.No.1 at 25 kg N/ha, if the cost of urea is Rs.12/kg

ANSWERS ▶

1. (a) N uptake by grain at 0 kg N/ha= 14.38 kg/ha
   N uptake by grain at 30 kg N/ha= 19.91 kg/ha
   N uptake by straw at 0 kg N/ha= 8.74 kg/ha
   N uptake by straw at 30 kg N/ha= 11.96 kg/ha

   (a) 5.47 kg/kg

   (b) 49.26 kg/kg

   (c) 18.74 kg/kg

   (d) 29.17 %

   (e) NHI at 0 kg N/ha= 0.62
   NHI at 30 kg N/ha= 0.62

   (f) 14.37 kg/kg

   (g) 77.25 kg/kg

   (h) 0.66 kg/kg

2. 1.16 kg seed/Rs. invested in N

WEED CONTROL

Weed control covers a vast area in Agronomy. The use of herbicide in crop and non-crop situation is increasing day by day. Besides, to judge the efficiency of herbicides as well as weed control methods, many indices are developed. In this chapter, an attempt has been made to discuss the numerical related to calibration of sprayers, requirement of herbicides and indices involved in control of weeds and weed survey.

7.1 CALIBRATION OF SPRAYERS

Spraying is the most common method of application of herbicides. Calibration of sprayer is pre-requisite for safe and effective control of weeds. A sub-lethal dose fails to give satisfactory weed control. On the other hand, overdose will affect the crop in crop land situation and will increase the costs in non-crop situation and will accumulate toxic residues in soil. Thus, calibration of sprayer and calculation of accurate amount of commercial product from the recommended dose is essential for effective control of weeds.

Example 1: Find out the volume of sprayer of water required to spray herbicide in 1 hectare land from the following details:

(a) Distance sprayed (length): 50 m

(b) Width of spray swath: 1.4 m

(c) Water required:2 lit.

Solution:

Area sprayed= 50 m × 1.4 m = 70 m²

70 m² area requires = 2 lit.

Therefore,10,000 m² area requires = $2 \times \dfrac{10{,}000}{70}$ = 285.7 litre

Example 2: Find out the volume of water required to spray herbicide in 1000 m² from the following details:

(a) Distance sprayed (length): 50 m

(b) Width of spray swath: 1.4 m

(c) Water required: 2 lit.

Solution:

$$\text{Spray volume} = \frac{\text{Area (m}^2) \times \text{Volume of water used (lit.)}}{\text{Width of Spray swath (m)} \times \text{Length covered (m)}}$$

Area sprayed= 50 m × 1.4 m = 70 m²

70 m² area requires = 2 lit.

Therefore,10,000 m² area requires = $2 \times \dfrac{1000}{70}$ = 28.57 litre

Example 3: Find out the volume of water required to spray herbicide in 2 hectare land from the following details:

(a) Distance sprayed (length): 50 m

(b) Width of spray swath: 1.4 m

(c) Water required: 2 lit.

Solution:

$$\text{Spray volume} = \frac{\text{Area (m}^2) \times \text{Volume of water used (lit.)}}{\text{Width of Spray swath (m)} \times \text{Length covered (m)}}$$

I hectare = 10,000 m²

Therefore, 2 hectare = 2 × 10,000 =20,000 m²

$$\text{Spray volume} = \frac{2 \times 20,000}{1.4 \times 50} = 571.4 \text{ litres}$$

Example 4: Find out the area sprayed by a full tank cf back pack sparyer from the following details:

(a)   Recommended spray volume: 400 lit./ha.

(b)   Volume of sprayer :12 lit.

Solution:

$$\textbf{Area sprayed } (\text{m}^2) = \frac{\text{Area (m}^2) \times \text{Volume of spray (lit.)}}{\text{Recommended spray volume (lit./ha)}}$$

400 lt. required = 10,000 m²

$$12 \text{ lt. required} = 12 \times \frac{10,000}{400} \text{ m}^2 = 300 \text{ m}^2$$

Therefore, area sprayed by a full tank back pack sprayer = 300 m²

Example 5: Find out the area sprayed by 5 tanks of back pack sparyer from the following details:

(a)   Recommended spray volume: 400 lit./ha.

(b)   Volume of sprayer :12 lit.

Solution:

$$\text{Area sprayed } (\text{m}^2) = \frac{\text{Area (m}^2) \times \text{Volume of spray (lit.)}}{\text{Recommended spray volume (lit./ha)}}$$

Volume of sprayer of I tank= 12 lit.

Therefore, volume of sparyer of 5 tanks= 12 × 5= 60 lit.

400 lt. required = 10,000 m^2

60 lt. required = $\dfrac{60 \times 10,000}{400}$ m^2 = 1500 m^2

Therefore, area sprayed by 5 tanks back pack sprayer = 1500 m^2

7.2 DOSE OF HERBICIDE

On the container or packet of the commercially available herbicide,concentration of the active ingredient is mentioned. For example, Hiltachlor 50 EC contain 50% Butachlor and Stomp 30 EC contains 30 % pendimethalin as emulsifying concentrate. As such, requirement of the commercial product has to be calculated to apply precise amount of herbicide.

$$\textbf{Dose of herbicide} = \frac{\text{Recommended rate of herbicide}}{\text{Active Ingredient}} \times 100$$

Example 6: Calculate the amount of Hiltachlor 50 EC to supply 1 kg butachlor

Solution:

Hiltachlor 50 EC, a commercial formulation contains 50% a.i. (active ingredient) of butachlor

Thus, to supply 50 kg butachlor, requirement of Hiltachlor 50 EC is = 100 lit.

Therefore, to supply 1 kg butachlor,requirement of Hiltachlor 50 EC is

$$= \frac{100}{50} = 2 \text{ litres}$$

Example 7: Calculate the amount of Hiltachlor 50 EC to apply butachlor @1.5 kg/ha in 3 ha.area

Solution:

$$\textbf{Dose of herbicide (kg or lit.)} = \frac{\text{Recommended rate (kg/ha)}}{\text{\% a.i. in herbicide formulation}} \times 100$$

Requirement of butachlor for 1 ha.= 1.5

Requirement of butachlor for 3 ha.= 1.5 × 3 =4.5 kg

Thus, to supply 50 kg butachlor, requirement of Hiltachlor 50 EC is = 100 lit.

Therefore, to supply 1 kg butachlor, requirement of Hiltachlor 50 EC is

$$= \frac{100}{50} = 2 \text{ litres}$$

Therefore, to supply 4.5 kg butachlor, requirement of Hiltachlor 50 EC is

$$= 2 \times 4.5 \text{ lit.} = 9.0 \text{ litres}$$

Hence, amount of Hiltachlor required for spray in 3 ha. area = 9.0 litres

Example 8: Calculate the amount of Hiltachlor 50 EC to apply butachlor @1.5 kg/ha in 5000 m^2 area.

Solution:

Dose of herbicide for 10,000 m^2 = 1.5 kg/ha

Therefore, dose of herbicide for 5000 $m^2 = \dfrac{1.5}{10,000} \times 5000 = 0.75$

Hence, dose of Hiltachlor for 5000 $m^2 = \dfrac{0.75}{50} \times 100 = 1.5$ lit.

Example 9: Glyphosate has to be applied in 30 cm band in sugarcane with 90 cm row spacing. The recommended rate is 1.6 kg/ha. How much Glycel 41 SL is needed to cover 5000 m^2?

Solution: Given that, herbicide is to be applied in 30 cm band with 90 cm spacing

Thus, $1/3^{rd}$ of the total area i.e. (5000/3) ha. will have to be sprayed with herbicide

Therefore, requirement of Glycel 41 SL = $1.6 \times \dfrac{0.5}{3} \times \dfrac{100}{41} = 0.65$ litre

Example 10: A farmer applied 1 lit. Stomp 30 EC (Pendimethalin) in his kabuli gram plot. What is the applicfation rate in terms of a.i./ha?

Solution:

Application rate of Stomp 30 EC= 1 lit./acre

1hectare = 2.5 acre

Therefore, application rate of Stomp 30 EC= 1 × 2.5=2.5 lit./ha.

1 lit. Stomp 30 EC=0.3 kg pendimethalin

Therefore,2.5 Lit Stomp 30 EC= 0.3 × 2.5 = 0.75 kg

Thus, application rate in terms of a.i. (pendimethalin)= 0.75 kg/ha.

Example 11: If 600 lit. spray water is required/ha, calculate the amount of water and the above herbicide (Example No.10) to be mixed per litre water.

Solution:

Water required for 10,000 m² = 600 lit.

Therefore,water required for 4000 m²= (600 × 4,000/10,000)= 240 litre

Hus. 1 lit. of Stomp 30 EcC has to be dissolved with 240 lit. of water

Therefore, Stomp 30 EC to be mixed/lit.water = $\dfrac{1}{240} \times \dfrac{1000}{240}$ ml = 4.16 ml

Hence, amount of water= 240 litre

Dose of Stomp = 4.16 ml.

Example 12: A student conducted an experiment on weed control with 4 treatments of butachlor at the rates of 0.75,1.30,1.25 and 1.5 kg a.i. with 5 replications. He has to apply Machetye 50 EC. If the plot size of the experiment is 6 m × 5 m,calculate the amount of herbicide used in the experiment.

Solution:

Plot size= 6.0 m × 5.0 m = 30 m^2 = 0.003 hectare

Requirement of Machete 50 EC/plot

For 0.75 kg a.i. /ha. $= \dfrac{0.75}{50} \times 100 = 1.5$ ml/ha.

For 0.003 hectare $= \dfrac{1.5}{10,000} \times 0.003 = 4.5$ ml./ha

For 1.00 kg a.i./ha. $= \dfrac{1}{50} \times 100 = 2$ ml/ha.

For 0.003 hectare $= \dfrac{2}{10,000} \times 0.003 = 6$ ml./ha

For 1.25 kg a.i. /ha. $= \dfrac{1.25}{50} \times 100 = 2.5$ ml/ha.

For 0.003 hectare $= \dfrac{2.5}{10,000} \times 0.003 = 7.5$ ml./ha

For 1.50 kg a.i./ha. $= \dfrac{1.50}{50} \times 100 = 3.0$ ml/ha.

For 0.003 hectare $= \dfrac{3.0}{10,000} \times 0.003 = 9.0$ ml./ha

Thus, total requirement of Machete 50EC/replication

= 4.5 +6.0 + 7.5+ 9.0 = 27.0 ml.

Therefore, total requirement of Machete 50 E for the experiment

= 27.0 ml × 5= 135.0 ml.

Example 13: A farmer is suggested to mix Weeder (Sulfosulfuron 75% WG) and Fuma (Fenoxaprop 10 EC) in a spray tank for weed control in 40 m × 30 m wheat plot. If the recommended rate of application of sulfosulfuron and fenoxaprop is 12.5 and 50 g/ha,respectively. Calculate the amount of herbicide for his wheat plot. Also, calculate the amount of herbicide to be mixed per tank of knapsack sprayer (tank capacity 15 litre), if 600 litre water is required to spray 1 ha. area.

Solution:

Area of wheat crop = 40 m × 30 m = 1200 m^2 = 0.12 ha.

Dose of weeder = $\dfrac{12.5}{75}$ × 100 = 16.6 kg/ha.

For 1200 m^2 = $\dfrac{16.6}{10,000}$ × 1200 = 1.9 ml./ha

Dose of Fuma = $\dfrac{50}{10}$ × 100 = 500 ml/ha.

For 1200 m^2 = $\dfrac{500}{10,000}$ × 1200 = 60 ml./ha

Requirement of water= $\dfrac{600}{10,000}$ × 1200 = 72 l/ha

Therefore, 1.9 ml of weeder and Fuma will have to be dissolved in 72 litres of water

Thus requirement of weeder per tank = 1.9 × $\dfrac{15}{72}$ = 0.39 ml

and requirement of fuma per tank = 60 × $\dfrac{15}{72}$ = 12.6 ml

As such, the weeder will have to mix 0.33 ml. Weeder and 12.6 ml. Fuma per tank for spraying hius wheat crop. Thus, 5 tank herbicide mixturte will be required for his crop.

Example 14: A farmer was suggested to apply 1% 2,4-D ethyl ester to kill the broad leaf weeds of his boundary wall. Calculate the amount of knockweed 36 EC required to be mixed per tank, if the capacity of sprayer tank is 12 litres.

Solution:

For 1% solution,100 lit. water requires = 1 lit. 2,4-D ethyl ester

Therefore, 12 lit. water requires = $\dfrac{12}{100}$ lit. 2,4-D ethyl ester

$$= 120 \text{ ml. 2,4-D ethyl ester}$$

Hence, requirement of knockweed 36 EC = $\dfrac{120}{36} \times 100 = 333.33$ ml.

Example 15: A farmer was suggested to apply 2% solution of 2,4-D to the ditch bank weeds. It is estimated that 1,000 lit. spray volume/ha will be required to wet the weeds completely. Calculate the amount of Fernoxone 80 WP required to treat 3 m wide and 2000 m long ditch tank.

Solution:

Area of ditch tank= 3 m × 2000= 6000 m^2

Requirement of spray volume= $1000 \times \dfrac{6000}{10,000} = 600$ lit.

Therefore, Fernoxone 80 WP requirement/lit. of water = $2 \times \dfrac{1000}{80} = 25$ g

Hence, Fernoxone 80 WP required 600 litre water

$$= 25 \times 600 = 15,000 \text{ g} = 15 \text{ kg}$$

Example 16: A farmer was suggested to treat his pond with 2 ppm 2,4-D ethyl ester to control submerged weeds. The length and width of the pond and depth of water are 50 m,20 m and 2.5 m,respectively. Calculate the amount of knockweed 36 EC required for the treatment of the pond.

Solution:

Volume of water $= 50 \times 20 \times 2.5$ m^3

$$= 2500 \text{ m}^3 = 25,00,000 \text{ lit.}$$

For 2 ppm solution =, 2 lit 2,4-D ethyl ester is needed in = 10,00,000 lit. water

Therefore, 25,00,000 lit. water will require $= 2 \times \dfrac{25,00,000}{10,00,000}$

$$= 5 \text{ lit. 2,4-D etyll ester}$$

Requirement of knockweed 36 EC$= 5 \times \dfrac{100}{36} = 13.89$ lit.

Example 17: Find out a.e. (acid equivalent) of 2,4-D dietyl amine salt if a commercial formulation which contains 70% a.i. of 2,4-D dietyl amine salt (Molecular weight of 2,4-D=221 and 2,4-D dietyl amine =326)

Solution:

Acid equivalent is the per cent original acid from which a salt or ester is derived.

Therefore, a.e. of 2,4-D dietyl amine $= \dfrac{221}{326} \times 100 = 67.79$ %

The formulation contains 70 % a.i.

Therefore a.e. of the commercial formulation $= \dfrac{70}{100} \times 67.79$

$$= 47.45 \text{ per cent}$$

Acid equivalent of 2,4-D dietyl amine = 47.45 per cent

7.3 COMMONLY USED INDICES IN WEED MANAGEMENT

There are many indices to judge the weed management practices in thedir ability to control weeds and realized higher yield.

7.3.1 Weed Control Efficiency (WCE)

It can be worked out taking into consideration the reduction in weed dry weight in treated plot over weed dry weight in unweeded check.

$$WCE = \frac{W_C - W_T}{W_C} \times 100$$

Where,

W_C = Weed dry weight in control (Unweeded) plot

W_T = Weed dry weight in treated plot

Example 18: If weed dry weight in weedy check plot and bispyribac treated plot in rice are 303 and 47 g/m^2 Calculate weed control efficiency.

Solution:

$$WCE = \frac{303 - 47}{47} \times 100 = 84.5 \%$$

7.3.2 Weed Control Index (WCI)

It can be worked out taking into consideration the reduction in weed population in treated plot over weed population in unweeded check.

$$WCI = \frac{WP_C - WP_T}{WP_C} \times 100$$

Where,

WP_C = Weed population in control (Unweeded) plot

WP_T = Weed population in treated plot

Example 19 : If weed population in weedy check plot and bispyribac treated plot in rice are 195.5 and 36.2/m^2 Calculate weed control index.

Solution:

$$WCI = \frac{195.5 - 36.2}{195.5} \times 100 = 81.5\ \%$$

7.3.3 Weed Smothering Efficiency (WSE)

It is worked out under intercropping situation.

$$WSE = \frac{W_S - W_T}{W_S} \times 100$$

Where,

W_S = Weed dry weight in sole crop plot

W_T = Weed dry weight in smother intercrop plot

Example 20: If weed dry weight in sole maize and maize + cowpea as smother crop plot are 178.3 and 105.6 g/m^2 Calculate weed smothering efficiency.

Solution:

$$WSE = \frac{178.3 - 105.6}{178.3} \times 100 = 40.8\ \%$$

7.3.4 Weed Index (WI)

It expresses yield loss due to weeds.

$$WI = \frac{Y_{WF} - Y_T}{Y_{WF}} \times 100$$

Where,

Y_{WF} = Yield in weed free plot

Y_T = Yield in treated plot

Example 21: If the grain yield of weed free plot is 26.2 q/ha and bispyribac treated plot is 18.7 q/ha. Calculate weed index.

Solution:

$$WI = \frac{26.2 - 18.7}{26.2} \times 100 = 28.62\%$$

7.3.5 Weed Persistence Index (WPI)

$$WPI = \frac{W_T}{W_C} \times \frac{WP_C}{WP_T}$$

Where,

W_C = Weed dry weight in control plot

W_T = Weed dry weight in treated plot

WP_C = Weed population in control plot

WP_T = Weed population in treated plot

Example 22 : Weed dry weight in bispyribac and weedy check plot in rice are 36.2 and 195.5 g/m², respectively. Corresponding weed population are 49 and 305/m². Calculate weed persistence index.

Solution:

$$\textbf{WPI} = \frac{36.2}{195.5} \times \frac{305}{49} = 1.15\%$$

7.3.6 Treatment Efficiency Index (TEI)

$$TEI = \frac{Y_T - Y_C}{Y_C} \div \frac{W_T}{W_C}$$

Where,

Y_T = Yield of treated plot

Y_C = Yield of control (unweeded) plot

W_C = Weed dry weight in control (unweeded) plot

W_T = Weed dry weight in treated plot

Example 23 : Weed dry weight in bispyribac and weedy check plot in rice are 36.2 and 195.5 g/m², respectively. Corresponding grain yield are 25.6 and 7.9 q/ha. Calculate treatment efficiency index.

Solution:

$$TEI = \frac{25.6 - 7.9}{7.9} \div \frac{36.2}{195.5} = 12.1$$

7.3.7 Weed Management Index (WMI)

$$WMI = \frac{Y_T - Y_C}{Y_C} \div \frac{W_C - W_T}{W_C}$$

Where,

Y_T = Yield of treated plot

Y_C = Yield of control (unweeded) plot

W_C = Weed dry weight in control (unweeded) plot

W_T = Weed dry weight in treated plot

Example 24 : Weed dry weight in bispyribac + hand weeding and weedy check plot in rice are 70 and 89.3 g/m², respectively. Corresponding grain yield are 21.16 and 13.53q/ha. Calculate weed management index.

Solution:

$$WMI = \frac{21.16 - 13.53}{13.53} \div \frac{89.3 - 70.0}{89.3} = 2.61$$

7.3.8 Agronomic Management Index (AMI)

$$AMI = \frac{Y_T - Y_C}{Y_C} - \frac{W_C - W_r}{W_C} \div \frac{W_C - W_r}{W_C}$$

Where,

Y_T = Yield of treated plot

Y_C = Yield of control (unweeded) plot

W_C = Weed dry weight in control (unweeded) plot

W_T = Weed dry weight in treated plot

Example 25 : Weed dry weight in bispyribac + hand weeding and weedy check plot in rice are 70 and 89.3 g/m^2, respectively.Corresponding grain yield are 21.16 and 13.53q/ha.Calculate agronomic management index.

Solution:

$$AMI = \frac{21.16 - 13.53}{13.53} - \frac{89.3 - 70}{89.3} \div \frac{89.3 - 70}{89.3} = 1.61$$

7.3.9 Integrated Weed Management Index (IWMI)

$$IWM\ I = \frac{WMI + AMI}{2}$$

Where,

WMI = Weed Management Index

AMI = Agronomic Management Index

Example 26 : Weed dry weight in bispyribac + hand weeding and weedy check plot in rice are 70 and 89.3 g/m^2, respectively.Corresponding grain yield are 21.16 and 13.53q/ha.Calculate integrated weed management index.

Solution:

Since,

WMI = 2.61

AMI = 1.61

$$IWMI = \frac{2.61 + 1.61}{2} = 2.11$$

7.3.10. Crop Resistance Index (CRI)

$$CRI = \frac{D_T}{D_C} \times \frac{W_C}{W_T}$$

Where,

D_T = Dry matter produced by the crop in treated plot

D_C = Dry matter produced by the crop in control (unweeded) plot

W_C = Weed dry weight in control (unweeded) plot

W_T = Weed dry weight in treated plot

Example 27 : Weed dry weight in bispyribac + hand weeding and weedy check plot in rice are 70 and 89.3 g/m^2, respectively. Corresponding crop dry weights are 635 and 89.3 g/m^2. Calculate crop resistance index.

Solution:

$$CRI = \frac{635}{368} \times \frac{89.3}{70} = 2.20$$

7.3.11. Weed Growth Rate (WGR)

$$WGR = \frac{W_2 - W_1}{t_2 - t_1}$$

Where,

W_1 = weed dry weight/area at time t_1

W_2 = weed dry weight/area at time t_2

Example 28 : Weed dry weight in weedy check plot in rice at 20 and 40 days after transplanting are 15.2 and 58.5 g/m^2, respectively. Calculate weed growth rate between 20-40 DAT.

Solution:

$$WGR = \frac{58.5\text{-}15.2}{40\text{-}20} = 2.17 \text{ g/m}^2$$

7.4 INDICES USED IN WEED SURVEY

Weed survey are useful for determining the occurrence and relative importance of weed species in crop production systems. Documenting the weed species present in different fields allow comparisons with past and future surveys. These competition can help to elucidate the effect of new weed control technologies on farming practices, document weed species shifts in response to new weed control technologies and to document the development of herbicide resistant weeds. Documenting the relative importance of weed species also facilitates the establishment of priorities for research and extension activities.

7.4.1 Frequency of Weeds

It is given by

$$F_k = \frac{\sum_{i=1}^{n} Y_i}{n} \times 100$$

Where

F_k = Frequency for species k

Y_i = Presence (I) or absence (0) of species k in field I

N = Number of fields surveyed

7.4.2 Field Uniformity (FU)

It is given by

$$FU_k = \frac{\sum_{i=1}^{n}\sum_{i=1}^{N}}{n.N} = Xij \times 100$$

where,

FU_k = Field uniformity of species k

X_{ij} = Presence (I) or absence (0) of species k in quadrate j in field I

N = Number of quadrates

7.4.3 Density

The density (D) of each species in a field is calculated by addition of number of plants in all quadrates and dividing by the area of total quadrates.

$$D_{ki} = \frac{\sum_{i=1}^{n} Z_j}{A_i}$$

Where,

D_{ki} = Density (in number/m^2) of species k in field I

Z_j = number of plants of a species in quadrat j

A_i = Area in m^2 of N quadrate in field I

7.4.4 Mean Field Density (MFD)

It is the mean number of plants/m^2 for each species averaged over all fields sampled.

$$MFD_k = \frac{\sum_{i=1}^{n} Zkj}{n}$$

where,

MFD_k = Mean field density of species k

Zkj = Density (in number/m^2) of species k in field j

N   = number of fields surveyed

7.4.5 Mean Occurrence Field Density (MOFD)

It is the number of plants/m^2 for a weed species averaged over only the fields in which that species occurs.

$$MFOD_k = \frac{\sum_{i=1}^{n} Dkj}{n - a}$$

Where,

$MFOD_k$ = Mean occurence field density of species k

Dki = Density (in number/m^2) of species k in field i

n = number of fields surveyed

a = Number of fields from which species k is absent

7.4.6 Relative Frequency (RF)

$$RF_k = \frac{\text{Frequency value of species } k}{\text{Sum of frequency values of all species}} \times 100$$

7.4.7 Relative Field Uniformity (RFU)

$$RFU_k = \frac{\text{Field Uniformity value of species } k}{\text{Sum of Field Uniformity values of all species}} \times 100$$

7.4.8. Relative Mean Field Density (RMFD)

$$RFD_k = \frac{\text{Mean Field Density value of species } k}{\text{Sum of Mean Field Density values of all species}} \times 100$$

7.4.9. Relative Abundance (RA)

The Relative Abundance of species k (RA_k) was calculated as the sum of Relative Frequency, Relative Field Uniformity and Relative Mean Field Density for that species

$$RA_k = RF_k + RFU_k + RMFD_k$$

A maximum Relative Abundance value of 300 would have been possible if only one species were found in all the fields that were surveyed.

Example 29: Calculate the Frequency, Field Uniformity, Density, Mean Field Density, Mean Occurrence Field Density, Relative Frequency, Relative Field Uniformity, Relative Mean Field Density and Relative Abundance of the following weed species surveyed in 5 fields with 10 quadrates (0.25 m²) in each field

					Number of weeds					
					Field 1					
					Quadrates					
Species	1	2	3	4	5	6	7	8	9	10
A	125	275	158	177	158	189	147	189	220	210
B	111	115	114	126	128	126	136	69	158	101
C	25	26	23	28	12	19	35	12	25	16
D	5	-	6	-	12	-	1	-	9	3
E	11	15	7	3	2	5	-	-	4	3
					Field 2					
A	108	105	126	155	126	128	125	139	108	129
B	255	272	189	205	260	178	205	269	158	202
C	5	15	12	25	22	15	13	18	25	12
D	6	-	-	1	-	-	2	1	3	-
E	11	15	6	4	2	-	-	5	6	3

Field 3

A	168	158	155	166	158	150	149	175	165	155
B	155	165	175	148	150	157	166	155	150	160
C	12	13	14	15	12	6	4	3	2	1
D	2	3	4	5	2	-	-	-	2	1
E	3	4	5	-	-	-	4	5	5	4

Field 4

A	135	270	159	187	160	195	159	175	200	205
B	98	90	105	111	105	103	95	98	101	89
C	12	11	15	18	19	24	15	12	13	14
D	-	-	5	-	6	-	-	5	8	2
E	3	-	3	3	-	5	-	-	5	-

Field 5

A	169	189	185	169	145	189	156	129	155	178
B	112	114	105	113	102	105	106	98	105	112
C	10	20	30	31	15	25	24	16	19	11
D	5	5	5	-	6	-	6	2	-	-
E	-	-	-	-	-	-	-	-	-	-

Solution

Number of weeds

Total

Field 1

Quadrates

Species	1	2	3	4	5	6	7	8	9	10	
A	125	275	158	177	158	189	147	189	220	210	1848
B	111	115	114	126	128	126	136	69	158	101	1184
C	25	26	23	28	12	19	35	12	25	16	221
D	5	0	6	0	12	0	1	0	9	3	36
E	11	15	7	3	2	5	0	0	4	3	50
Sub-total	277	431	308	334	312	339	319	270	416	333	3339

Field 2										
A	108	105	126	155	126	128	125	139	108	129	1249
B	255	272	189	205	260	178	205	269	158	202	2193
C	5	15	12	25	22	15	13	18	25	12	162
D	6	0	0	1	0	0	2	1	3	0	13
E	11	15	6	4	2	0	0	5	6	3	52
Sub-total	385	407	333	390	410	321	345	432	300	346	3669

Field 3										
A	168	158	155	166	158	150	149	175	165	155	1599
B	155	165	175	148	150	157	166	155	150	160	1581
C	12	13	14	15	12	6	4	3	2	1	82
D	2	3	4	5	2	0	0	0	2	1	19
E	3	4	5	0	0	0	4	6	5	4	31
Sub-total	340	343	353	334	322	313	323	339	324	321	3312

Field 4										
A	135	270	159	187	160	195	159	175	200	205	1845
B	98	90	105	111	105	103	95	98	101	89	995
C	12	11	15	18	19	24	15	12	13	14	153
D	0	0	5	0	6	0	0	5	8	2	26
E	3	0	3	3	0	5	0	0	5	0	19
Sub-total	248	371	287	319	290	327	269	290	327	310	3038

Field 5										
A	169	189	185	169	145	189	156	129	155	178	1664
B	112	114	105	113	102	105	106	98	105	112	1072
C	10	20	30	31	15	25	24	16	19	11	201
D	5	5	5	0	6	0	6	2	0	0	29
E	0	0	0	0	0	0	0	0	0	0	0
Sub-total	296	328	325	313	268	319	292	245	279	301	2966
Grand total	1546	1880	1606	1690	1602	1619	1548	1576	1646	1611	16324

1. (a) Frequency of weed species A, FA $= 5 \times \dfrac{100}{5} = 100$

    (b) Frequency of weed species B, FB $= 5 \times \dfrac{100}{5} = 100$

    (c) Frequency of weed species C, FC $= 5 \times \dfrac{100}{5} = 100$

    (d) Frequency of weed species D, FD $= 5 \times \dfrac{100}{5} = 100$

    (e) Frequency of weed species E, FE $= 4 \times \dfrac{100}{5} = 80$

2. (a) Field Uniformity of weed species A, FA

    $$= (10 + 10 + 10 + 10 + 10) \times \dfrac{100}{50} = 100$$

    (b) Field Uniformity of weed species B, FB

    $$= (10 + 10 + 10 + 10 + 10) \times \dfrac{100}{50} = 100$$

    (c) Field Uniformity of weed species C, FC

    $$= (10 + 10 + 10 + 10 + 10) \times \dfrac{100}{50} = 100$$

    (d) Field Uniformity of weed species D, FD

    $$= (6 + 5 + 7 + 5 + 6) \times \dfrac{100}{50} = 58$$

    (e) Field Uniformity of weed species E, FE

    $$= (8 + 8 + 7 - 5 + 10) \times \dfrac{100}{50} 0 = 56$$

3. Density:

In Field 1, Density of species A, DA1 = $\dfrac{1848}{10 \times 0.25}$ = 739.2 nos./m^2

In Field 1, Density of species B, DB1 = $\dfrac{1184}{10 \times 0.25}$ = 473.6 nos./m^2

In Field 1, Density of species C, DC1 = $\dfrac{221}{10 \times 0.25}$ = 88.4nos./m^2

In Field 1, Density of species D, DD1 = $\dfrac{36}{10 \times 0.25}$ = 14.4 nos./m^2

In Field 1, Density of species E, DE1 = $\dfrac{50}{10 \times 0.25}$ = 20.0 nos./m^2

In Field 2, Density of species A, DA2 = $\dfrac{1249}{10 \times 0.25}$ = 499.6 nos./m^2

In Field 2, Density of species B, DB2 = $\dfrac{2193}{10 \times 0.25}$ = 877.2 nos./m^2

In Field 2, Density of species C, DC2 = $\dfrac{162}{10 \times 0.25}$ = 64.8nos./m^2

In Field 2, Density of species D, DD2 = $\dfrac{13}{10 \times 0.25}$ = 5.2 nos./m^2

In Field 2, Density of species E, DE2 = $\dfrac{52}{10 \times 0.25}$ = 20.8 nos./m^2

In Field 3, Density of species A, DA3 = $\dfrac{1599}{10 \times 0.25}$ = 639.6 nos./m^2

In Field 3, Density of species B, DB3 = $\dfrac{1581}{10 \times 0.25}$ = 632.4 nos./m2

In Field 3, Density of species C, DC3 $= \dfrac{81}{10 \times 0.25} = 32.8 \text{nos.}/\text{m}^2$

In Field 3, Density of species D, DD3 $= \dfrac{19}{10 \times 0.25} = 7.6 \text{ nos.}/\text{m}^2$

In Field 3, Density of species E, DE3 $= \dfrac{31}{10 \times 0.25} = 12.4 \text{ nos.}/\text{m}^2$

In Field 4, Density of species A, DA4 $= \dfrac{1845}{10 \times 0.25} = 738.0 \text{ nos.}/\text{m}^2$

In Field 4, Density of species B, DB4 $= \dfrac{995}{10 \times 0.25} = 398.0 \text{ nos.}/\text{m}^2$

In Field 4, Density of species C, DC4 $= \dfrac{153}{10 \times 0.25} = 61.2 \text{ nos.}/\text{m}^2$

In Field 4, Density of species D, DD4 $= \dfrac{26}{10 \times 0.25} = 10.4 \text{ nos.}/\text{m}^2$

In Field 4, Density of species E, DE4 $= \dfrac{19}{10 \times 0.25} = 7.6 \text{ nos.}/\text{m}^2$

In Field 5, Density of species A, DA5 $= \dfrac{1664}{10 \times 0.25} = 665.5 \text{ nos.}/\text{m}^2$

In Field 5, Density of species B, DB5 $= \dfrac{1072}{10 \times 0.25} = 428.8 \text{ nos.}/\text{m}^2$

In Field 5, Density of species C, DC5 $= \dfrac{201}{10 \times 0.25} = 80.4 \text{ nos.}/\text{m}^2$

In Field 5, Density of species D, DD5 $= \dfrac{29}{10 \times 0.25} = 11.6 \text{ nos.}/\text{m}^2$

In Field 5, Density of species E, DE5 $= \dfrac{0}{10 \times 0.25} = 0.0 \text{ nos.}/\text{m}^2$

4. Mean Field Density

Mean Field Density of Species A, MFD_A

$$= \frac{739.2 + 499.6 + 639.6 + 738.0 + 665.5}{5} = 656.38 \text{ nos./m}^2$$

Mean Field Density of Species B, MFD_B

$$= \frac{473.6 + 877.2 + 632.4 + 98.0 + 428.8}{5} = 562.0 \text{ nos./m}^2$$

Mean Field Density of Species C, MFD_C

$$= \frac{473.6 + 877.2 + 632.4 + 98.0 + 428.8}{5} = 562.0 \text{ nos./m}^2$$

Mean Field Density of Species D, MFD_D

$$= \frac{14.4 + 5.2 + 7.6 + 10.4 + 11.6}{5} = 8.84 \text{ nos./m}^2$$

Mean Field Density of Species E, MFD_E

$$= \frac{20.0 + 20.8 + 12.4 + 7.6 + 0}{5} = 12.16 \text{ nos./m}^2$$

5. Mean Occurrence Field Density

Mean Occurrence Field Density of Species A, $MOFD_A$

$$= \frac{739.2 + 499.6 + 639.6 + 738.0 + 665.5}{5} = 656.38 \text{ nos./m}^2$$

Mean Occurrence Field Density of Species B, $MOFD_B$

$$= \frac{473.6 + 877.2 + 632.4 + 98.0 + 428.8}{5} = 562.0 \text{ nos./m}^2$$

Mean Occurrence Field Density of Species C, $MOFD_C$

$$= \frac{473.6 + 877.2 + 632.4 + 98.0 + 428.8}{5} = 562.0 \text{ nos./m}^2$$

Mean Occurrence Field Density of Species D, $MOFD_D$

$$= \frac{14.4 + 5.2 + 7.6 + 10.4 + 11.6}{5} = 8.84 \text{ nos./m}^2$$

Mean Occurrence Field Density of Species E, $MOFD_E$

$$= \frac{20.0+ +20.8 + 12.4 + 7.6}{4} = 15.2 \text{ nos./m}^2$$

6. Relative Frequency

Relative Frequency of Species A, RF_A

$$= 100 \times \frac{100}{100+100+100+100+80} = 20.83$$

Relative Frequency of Species B, RF_B

$$= 100 \times \frac{100}{100+100+100+100+80} = 20.83$$

Relative Frequency of Species C, RF_C

$$= 100 \times \frac{100}{100+100+100+100+80} = 20.83$$

Relative Frequency of Species D, RF_D

$$= 100 \times \frac{100}{100+100+100+100+80} = 20.83$$

Relative Frequency of Species E, RF_E

$$= 80 \times \frac{100}{100+100+100+100+80} = 16.67$$

7. Relative Field Uniformity:

Relative Field Uniformity of Species A, RFU_A

$$= 100 \times \frac{100}{100+100+100+58 +56} = 24.15$$

Relative Field Uniformity of Species B, RFU_B

$$= 100 \times \frac{100}{100+100+100+58+56} = 24.15$$

Relative Field Uniformity of Species C, RFU_C

$$= 100 \times \frac{100}{100+100+100+58+56} = 24.15$$

Relative Field Uniformity of Species D, RFU_D

$$= 58 \times \frac{100}{100+100+100+58+56} = 14.01$$

Relative Field Uniformity of Species E, RFU_E

$$= 56 \times \frac{100}{100+100+100+58+56} = 13.53$$

8. Relative Mean Field Density

Relative Mean Field Density of Species A, $RMFD_A$

$$= 656.38 \times \frac{100}{656.38+562.0+65.52+8.84+12.16} = 50.30$$

Relative Mean Field Density of Species B, $RMFD_B$

$$= 562 \times \frac{100}{656.38+562.0+65.52+8.84+12.16} = 43.07$$

Relative Mean Field Density of Species C, $RMFD_C$

$$= 65.52 \times \frac{100}{656.38+562.0+65.52+8.84+12.16} = 5.02$$

Relative Mean Field Density of Species D, $RMFD_D$

$$= 8.84 \times \frac{100}{656.38+562.0+65.52+8.84+12.16} = 0.68$$

Relative Mean Field Density of Species E, $RMFD_E$

$$= 12.16 \times \frac{100}{656.38 + 562.0 + 65.52 + 8.84 + 12.16} = 0.93$$

9. Relative Abundance

Relative Abundance of Species A, $RA_A = 20.83 + 24.15 + 50.30 = 95.28$

Relative Abundance of Species B, $RA_B = 20.83 + 24.15 + 43.07 = 88.05$

Relative Abundance of Species C, $RA_C = 20.83 + 24.15 + 5.02 = 50.00$

Relative Abundance of Species D, $RA_D = 20.83 + 14.01 + 0.68 = 35.52$

Relative Abundance of Species E, $RA_E = 16.67 + 13.53 + 0.93 = 31.13$

⋮ EXERCISES ▶

1. Find out the volume of water required to spray herbicide in 10 ha. land from the following details:

    (a) Distance sprayed (length)= 100 cm

    (b) Width of spray swath= 1.4 m

    (c) Water required = 4.5 lit.

2. Find out the volume of water required to spray herbicide in 3000m² land from the following details:

    (a) Distance sprayed (length)= 100 cm

    (b) Width of spray swath= 1.5 m

    (c) Water required = 4.5 lit.

3. Find out the area sprayed by 10 tank of knapsack sprayer from the following details:

    (a) Recommended spray volume= 600 lit./ha

    (b) Volume of the sprayer = 16 lit.

4. Calculate the amount of Machete50 EC to apply Butachlor at the rate of 1kg/ ha in 5000m² and 5 ha. area

5. A herbicide formulation contains active ingredient of 250g/lit. and desired rate of application is 1.5 kg/ha. Calculate the volume of herbicide formulation required for 2.5 ha. area.

6. A farmer want to apply paraquat in 30 cm band with 90 cm row spacing. The recommended rate of application is 1.6 kg/ha. How much farmer need Uniquat 24 SL to cover one hectare.

7. Calculate the amount of Basalin 45 EC required to apply in greengram from the following details:

   (a) Plot size = 50 m × 40 m

   (b) Recommended rate = 0.75 kg/ha (Fluchloralin)

8. If Basalin 45 EC is not available, instead ,treflan 48 EC (Trifluralin) which is to be applied at 1 kg /ha is available in the market. Find out the amount of Treflan 48 EC the farmer has to apply in his greengram plot.

9. A farmer purchased 5 litre Stomp 30 EC for his greengram crop. If the recommended rate of pendimethalin is 0.75 kg/ha. Calculate how much area he can cover with the purchased herbicide.

10. Find out the volume of water,herbicide and herbicide/lit. to spray herbicide in 0.5 ha. blackgram from the following details:

    (a) 6 lit. water is required to spary 100 m² area

    (b) Herbicide= stomp 30 EC

    (c) Recommended rate= 0.75 kg/ha

11. How much area can be covered with 10 Kg Machete 5G granule,if applied at 1.0 kg/ha.

12. A student conducted an experiment on weed control with 6 treatments of butachlor @ 0.75,1.0,1.25,1.5, 1.75 and 2.0 kg/ha. with 3 replications. He used Machete 10 G. If the plot size of the experiment is 6m × 5m,calculate the amount of herbicide used in the experiment.

13. A farmer is suggested to mix Weeder (Sulfosulfuron) 75% WG) and Axial XL 100 EC (Pinoxaden) in a spray tabk for weed control in

one acre wheat plot. If the recommended rate of sulfosulfuron and pinoxaden is 12.5 and 25 g/ha, respectively. Calculate the amount of herbicide for his wheat plot. Also calculate the amount of herbicide to be mixed per tank of knapsack sprayer (tank capacity= 16 lit.), if 600lit. water is required to spray 1 ha, area.

14. A farmer was suggested to treat his pond with 2 ppm 2,4_D ethyl ester to control the submerged weeds. The length and width of the pond and depth of water are 50 m, 20 m and 2.5 m, respectively. The farmer had wrongly calculated the herbicide. Instead of 2 ppm, it was 1.25 ppm. Calculate the additional amount of knockweed 36 EC required to make the concentration of 2,4- D ethyl ester 2 ppm.

15. Find out the quantity of knockweed 36EC and water required to spray 10 ha. wheat crop with a solution of 600 ppm 2,4_d. The recommended dose is 0.50 kg/ha.

16. Find out the quantity of Fernoxone 80 WP and water required to spray 5 ha. wheat crop with a solution of 1000 ppm 2,4-D. the recommended dose is 0.50 kg/ha.

17. How much Machete 50 EC will be required for 1000 m², 1 ha. and 1.5 ha. rice area, if butachlor is to be applied as pre-emergence at the rate of 1.5 kg/ha.

18. Find out a.e. (acid equivalent) of ester of 2,4- D if a commercial formulation which contains 50% a.i. (Molecular weight of 2,4-D = 221 and 2,4- D ester =242) of 2,4-ester.

19. Calculate the amount of herbicide required for an area of 2500 m², if 2,4-D Na salt is to be applied as post emergence at the rate of 0.69 a.e./ha

20. Calculate the amount of Butachlor 50 EC required for an area of 1000 m², if butachlor is to applied as pre-emergence at the rate of 0.2% concentration dissolved in 800 lit. water/ha for the application of the herbicide.

21. Calculate the concentration of 2,4-D ethyl ester in ppm if 10 ml. knockweed 36 EC (which contains 36% 2,4-D ethyl ester) dissolved in 10 lit. water.

22. It is suggested tp apply atrazine at the rate of 1 kg/ha as pre-emergence application in a sugarcane plot of 100 m × 45 m where sugarcane is planted along the length with row to row distance 90 cm. Calculate the quantity of Markanzine 50 WP required, if

    (a)   the herbicide is applied in entire plot

    (b)   the herbicide is planted at 30 cm band only within planted rows

23. If weed dry weight in weedy check plot and weed free plot in rice are 350.6 and 40.3 g/m^2. Calculate Weed Control Index.

24. If weed dry weight in sole maize and maize + ricebean are 200.6 and 103.6 g/m^2. Calculate Weed Smothering Efficiency.

25. Weed dry weight in bispyribac and weedy check plot in rice are 42.6 and 212.9 g/m^2 respectively. Corresponding weed population and grain yield are 56.6 and $325.9/m^2$ and 26.7 and 11.3 q/ha. Calculate Weed Control Index, Weed Control Efficiency, Weed Persistence Index, Treatment Efficiency Index, Weed Management Index, Agronomic Management Index and Integrated Weed Management Index.

26. Weed dry weight in weedy check plot and bispyribac + hand weeding in rice are 156.8 and 46.8 g/m^2 respectively. Corresponding crop dry weight are 383 and 655 g/m^2. Calculate Crop Resistance Index.

27. Weed dry weight in weedy check plot in transplanted rice at 15 and 30 DAT are 18.2 and 55.8 g/m^2 respectively. Calculate Weed Growth Rate between 15-30 DAT.

28. Calculate Frequency, Field Uniformity,Density, Mean Field Density, Mean Occurrence Field Density, Relative Frequency,Relative Field Uniformity,Relative Mean Field Density and Relative Abundance of the following weed species surveyed in 3 fields with 10 quadrates (0.25 m^2) in each field.

Number of Weeds

Field 1

Quadrates

Species	1	2	3	4	5	6	7	8	9	10
A	130	270	165	180	165	199	159	180	200	201
B	99	105	108	126	115	126	120	89	126	105
C	20	12	15	28	19	18	26	18	15	19
D	4	2	-	-	-	-	8	6	9	3
E	7	-	7	3	2	5	-	-	4	3

Field 2

A	101	105	126	124	126	128	125	139	108	129
B	255	205	189	205	157	166	155	150	160	202
C	12	11	10	21	19	18	15	15	21	19
D	3	4	-	1	-	-	2	1	3	-
E	-	-	-	-	-	-	-	-	-	-

Field 3

A	150	186	155	166	158	150	149	175	165	155
B	159	181	175	148	166	157	166	155	150	160
C	12	11	-	11	12	6	2	-	2	1
D	2	3	4	5	2	-	-	-	2	1
E	-	-	-	-	-	-	-	-	-	-

⋮ ANSWERS ▶

1. 3214 lit.
2. 90 lit.
3. 2667 lit.
4. (a)1 lit. (b)10 lit.
5. 15 lit.
6. 2.22 lit.
7. 0.333 lit.
8. 0.417 lit.
9. 2 ha.
10. Volume of water=300lit.; Herbicide=1.25 lit.; Herbicide= 4.17 ml/litre
11. 0.5 ha
12. 742.5 g

13. Weeder=6.67 ml/acre; Axial XL= 10 ml./acre; Weeder= 0.1779 ml/tank;Axial XL= 0.2667 ml/tank

14. 5.21 lit.   **15.**6.94 lit;4166.7 lit.   **16.** 3.12kg;2500 lit.

17. 0.3 lit.;3lit.;4.5lit.

18. 45.66%   **19.** 0.31 kg   **20.** 320 ml

21. 360 ppm   **22.** (a) 0.90 kg (b) 0.30kg **23.** 88.51%

24. 48.35 %

25. Weed control Index= 79.57 %; Weed Control Efficiency= 82.63%;Weed Persistence index=1.15;Treatment Efficiency Index=6.81;Weed Management Index=1.70;Agronomic Management Index=0.70;Integrated Weed Management Index= 1.20

26. 5.73   **27.** 2.51g/m^2/day

28.

Species	Density Field 1	Field 2	Field 3	F	FU	MFD	MOFD	RF	RFU	RMRDD	RA
A	739.6	484.4	643.6	100	100	622.5	622.5	23.1	26.3	47.9	97.3
B	447.6	737.6	646.8	100	100	610.7	610.7	23.1	26.3	47.0	96.4
C	76.0	64.4	22.8	100	93.3	54.4	54.4	23.1	24.6	4.2	51.0
D	12.8	5.6	7.6	100	63.3	8.7	8.7	23.1	16.7	0.7	40.5
E	12.4	0	0	33.3	23.3	401	12.4	7.7	6.1	0.3	14.1

ASSESSMENT OF LAND USE AND YIELD ADVANTAGE

Global population is increasing day by day. However, Land resources are limited and to feed this increasing population, agricultural lands will have to be utilized to the best of its capacities. The only way to increase agricultural production in the small or marginal units of farming is to increase the productivity per unit time and area. This may be achieved by several ways, such as, by breeding more productive varieties or quicker maturing varieties with equal yields or by improving managements practices like fertilizer use, weed and pest control etc. the other factor to be added is multiple cropping. Multiple cropping is a practice of getting maximum production from a unit area in a certain time span. It is a philosophy of maximum crop production per unit area of land and can be achieved by by crop rotations, relay cropping, mixed cropping, inter cropping etc.

Large number of indices are available for assessing land use and the yield advantage of crop mixtures compared to pure stands. These indices not only reflect the differences in criteria used to appraise "often encompassing aspects of quality or value as well as yield, but also reflect the different reasons for which an assessment is made.

8.1. ROTATIONAL INTENSITY

It is calculated by-

$$\text{Rotational Intensity} = \frac{\text{Number of crops grown in rotation}}{\text{Duration of rotation (year)}} \times 100\%$$

Example 1. Calculate the rotational intensity of the following crop sequences

Rice- wheat	(1 year)
Rice- Early Potato-Wheat-Cowpea	(1 year)
Sesamum- Greengram-Broccoli-Blackgram-Greengram-Cauliflower	(2 year)
Maize-Potato-Sugarcane-Ratoon-Greengram	(3 year)
Maize-Sugarcane+ Potato-Ratoon+Mustard-Greengram	(3 year)

Solution:

(a) Rotational intensity $= \dfrac{2}{1} \times 100\% = 200\%$

(b) Rotational Intensity $= \dfrac{4}{1} \times 100\% = 100\%$

(c) Rotational Intensity $= \dfrac{6}{2} \times 100\% = 300\%$

(d) Rotational Intensity $= \dfrac{5}{3} \times 100\% = 167\%$

8.2. Cropping Intensity:

It is calculated by-

$$\text{Rotational Intensity} = \frac{\text{Gross cropped area}}{\text{Net cultivated area}} \times 100\%$$

Example 2: Calculate the cropping intensity from the following information:

(a) Net cultivated area of the farm = 10ha

     Gross cultivated area   = 15ha

(a) Net cultivated area   = 10ha

     Only kharif rice is grown in entire area and thereafter the land is kept fellow

(b) Net cultivated area   = 10ha

     Kharif rice-rapeseed sequence is followed in entire area

Solution:

(a) Cropping Intensity $\quad = \dfrac{15}{10} \times 100\%$

$\qquad\qquad\qquad\qquad = 150\%$

(b) Cropping Intensity $\quad = \dfrac{10}{10} \times 100\%$

$\qquad\qquad\qquad\qquad = 100\ \%$

(c) Cropping Intensity $\quad = \dfrac{20}{10} \times 100\%$

$\qquad\qquad\qquad\qquad = 200\ \%$

Example 3: Calculate the cropping intensity of a 20 ha farm where land is utilized as fellows:

Block	Area(ha)	Sequence		
		Summer	Kharif	Rabi
A	5	-	Rice	Mustard
B	5	-	Rice	Rice
C	4	-	Rice	Wheat
D	1	-	Rice	Tomato
E	2	Greengram	Baby corn	Potato
F	2	Blackgram	Baby corn	Knolkhol
G	1	Okra	Greengram	Broccoli

Solution:

Name of the crop	Area(ha)
Rice(Block A)	5.0
Mustard(Block A)	5.0
Rice(Block B)	5.0
Rice(Block B)	5.0
Rice(Block C)	4.0
Wheat (Block C)	4.0
Tomato(Block D)	1.0
Rice(Block D)	1.0
Greengram(Block E)	2.0
Baby cron(Block E)	2.0
Potato(Block E)	2.0
Blackgram(Block F)	2.0
Baby cron(Block F)	2.0
Knolkhol(BlockF)	2.0
Okra(Block G)	1.0
Greengram(Block G)	1.0
Broccoli(Block G)	1.0
Gross cropped area	45.0

$$\text{Cropping intensity} = \frac{45}{20} \times 100$$

$$= 225 \%$$

8.3. MULTIPLE CROPPING INDEX (MCI)

It is similar to Cropping intensity (CI). It measures the sum of area under varius crops raised in single year divided by net area available for that cropping pattern and expressed in percentage. It is generally calculated for each cropping pattern separately.

Thus, it is calculated by-

$$MCI = \frac{\sum\limits_{i=1}^{n} ai}{A} \times 100\%$$

Where,

I = 1,2,3,………………….n

N = total number of crops

Ai = area occupied by the ith crop,

A = total land area available for cultivation

[Do you have any confusion on CI and MCI ? $\sum\limits_{i=1}^{n} ai$ Isn't = Gross cropped area and A= Net cropped area ?]

Example 4. Calculate MCI for each cropping sequence and entire farm of Example 3.

Solution:

MCI for Block A $= \frac{10}{5} \times 100\% = 200\%$

MCI for Block B $= \frac{10}{5} \times 100\% = 200\%$

MCI for Block C $= \frac{10}{5} \times 100\% = 200\%$

MCI for Block D $= \frac{10}{5} \times 100\% = 200\%$

MCI for Block E $= \frac{6}{2} \times 100\% = 300\%$

MCI for Block F $\quad = \dfrac{6}{2} \times 100\% = 300\%$

MCI for Block G $\quad = \dfrac{3}{1} \times 100\% = 300\%$

MCI for wholfarm $\quad = \dfrac{45}{20} \times 100\% = 225\%$

Whole farm $\quad = \dfrac{45}{20} \times 100\% = 225\%$

8.4 LAND USE EFFICIENCY

land use efficiency indicates the percent land use in term of duration. This, if LAU is 70%, it indicates the out of 100 days, land was occupied by crop for 70 days. LUE is given by-

$$LUE = \dfrac{\sum\limits_{i=1}^{n} D_i}{365} \times 100$$

Where,

I $\quad$ = 1,2,3,………………….n

N $\quad$ = total number of crops

Di $\quad$ = Number of days occupied by the ith crop,

Example 5: Following crops are grown in 2 fields. Calculate the LAU

Block	Sequence		
	Summer	Kharif	Rabi
I	-	Rice(120)*	Mustard(120)
II	-	Rice(120)	Rice(130)

*Duration occupied by the crop in the field

Solution:

LUE for field I $\quad = \dfrac{120+120}{365} \times 100\% = 65.75\%$

LUE for field I I $\quad = \dfrac{120+130}{365} \times 100\% = 68.49\%$

8.5. CULTIVATED LAND UTILIZATION INDEX (CLUI)

Cultivated Land Utilization Index (CLUI) is calculated by summing the products of land area to each crop, multiplied by the Actual cultivated land times 365 days.

$$CLUI = \dfrac{\sum\limits_{i=1}^{n} a_i d_i}{A \times 365}$$

Where,

I $\quad = 1,2,3,\dots\dots\dots\dots\dots.n$

n $\quad$ = total number of crops

a_i $\quad$ = area occupied by ith crop

d_i $\quad$ = days that the ith crop occupied

A $\quad$ = total cultivated land area(available for 365 days)

CLUI can be expressed as fraction or percentage. This gives an idea about how the land area has been put into use. If the index is less than 1 (i.e. 100%), it shows that the land is been left fellow.

Example 6. A farmer has 23 ha cultivable land available for entire year. The land is utilized as follow:

Block	Area(ha)	Sequence		
		Summer	**Kharif**	**Rabi**
A	5	-	Rice(120)*	Mustard(120)
B	5	-	Rice(120)	Rice(130)
C	4	-	Rice(120)	Wheat(110)
D	1	-	Rice(120)	Tomato(110)
E	2	Greengram(85)	Baby corn(70)	Potato(120)
F	2	Blackgram(75)	Baby corn(70)	Knolkhol(50)
G	1	Okra(120)	Greengram(70)	Broccoli(70)
H	3	Vegetable(120)	-	Rapeseed(95)

*duration occupied by the crop in the field

Calculate CLUI for each block and for the entire farm

Solution:

$$\text{CLUI(for block A)} = \frac{5 \times 120 + 5 \times 120}{5 \times 365} = 0.66$$

$$\text{CLUI(for block B)} = \frac{5 \times 120 + 5 \times 130}{5 \times 365} = 0.68$$

$$\text{CLUI(for block C)} = \frac{4 \times 120 + 4 \times 110}{4 \times 365} = 0.63$$

$$\text{CLUI(for block D)} = \frac{1 \times 120 + 1 \times 110}{1 \times 365} = 0.63$$

$$\text{CLUI(for block E)} = \frac{2 \times 85 + 2 \times 70 + 2 \times 120}{2 \times 365} = 0.75$$

$$\text{CLUI(for block F)} = \frac{2 \times 75 + 2 \times 70 \times 50}{2 \times 365} = 0.53$$

$$\text{CLUI(for block G)} = \frac{1 \times 120 + 1 \times 70 + 1 \times 70}{1 \times 365} = 0.71$$

$$\text{CLUI(for block H)} = \frac{3 \times 120 + 3 \times 95}{3 \times 365} = 0.59$$

8.6. CROPPING INTENSITY INDEX (CII)

Cropping intensity index (CII) assesses farmer's actual land use in area and time relationship for each crop or group of crops compared to the total available land area and time, including land thart is temporarily available for cultivation.

$$CII = \frac{\sum_{i=1}^{n} a_i t_i}{A_0 T + \sum_{j=1}^{m} a_i t_i}$$

Where,

i   = 1,2,3,...........................n

j   = 1,2,3,...........................m

n   = total number of crops grown during the time period T(usually 365 days)

m   = total number of fields temporarily available

a_i = area occupied by ith crop

t_i = duration of the ith crop

A_0 = cultivated land area of the farmer for entire period T

A_j = land area of jth field

Tj = time period Aj is available

Example 7. A farmer has 20ha cultivable land available for entire year. In addition, he has 3 ha flood prone Char land which is temporarily available to him for cultivation from October to May(243 days). The land utilized as follow:

Block	Area(ha)	Sequence		
		Summer	**Kharif**	**Rabi**
A	5	-	Rice(120)*	Mustard(120)
B	5	-	Rice(120)	Rice(130)
C	4	-	Rice(120)	Wheat(110)
D	1	-	Rice(120)	Tomato(110)
E	2	Greengram(85)	Baby corn(70)	Potato(120)
F	2	Blackgram(75)	Baby corn(70)	Knolkhol(50)
G	1	Okra(120)	Greengram(70)	Broccoli(70)
H	3	Vegetable(120)	-	Rapeseed(95)

*Duration occupied by the crop in the field[in some literature, it is expressed on month. When put in the formula, duration willo ultimately be cancelled and result will be same. Thus it does not matter wheather it is in days or months.]

Calculate CII of the entire farm.

Solution:

$$CII = \frac{\begin{array}{c}600+600+600+650+480+440+120+110+170+140\\+240+150+140+100+120+70+70+[3\times120+3\times95]\end{array}}{20\times365+3\times243} = \frac{5445}{8029} = 0.68$$

8.7. SPECIFIC CROP INTENSITY INDEX SCII)

It is derivative to CII. It determines the amount of area time denotes to each crop or group of crops compared to the total time available to the farmer.

$$SCII = \frac{\sum\limits_{k=1}^{NK} a_k T_i}{A_0 T + \sum\limits_{j=1}^{m} A_i T_i}$$

Where,

k = 1,2,3,...........................N_k

j = 1,2,3,...........................m

N_k = total number of crops within aspecific designation such as vegetable or field crops grown by the farmers during the time period T

m = total number of fields temporarily available

a_k = area occupied by kth crop

t_k = duration of the kth crop

A_0 = cultivated land area of the farmer for entire period T

A_j = land area of jth field

Tj = time period Aj is available

Example 8. A farmer has 20ha cultivable land available for entire year. In addition, he has 3 ha flood prone Char land which is temporarily available to him for cultivation from October to May (243 days). The land utilized as follow:

Block	Area(ha)	Sequence		
		Summer	Kharif	Rabi
A	5	-	Rice(120)*	Mustard(120)
B	5	-	Rice(120)	Rice(130)
C	4	-	Rice(120)	Wheat(110)
D	1	-	Rice(120)	Tomato(110)
E	2	Greengram(85)	Baby corn(70)	Potato(120)
F	2	Blackgram(75)	Baby corn(70)	Knolkhol(50)
G	1	Okra(120)	Greengram(70)	Broccoli(70)
Char Land	3	Vegetable(120)	-	Rapeseed(95)

*duration occupied by the crop in the field

117

Solution:

SCII for cereals

$$= \frac{120\times5+120\times5+130\times5+120\times4+110\times4+120\times1+120}{20\times365+3\times243} = \frac{2890}{8029} = 0.36$$

$$\text{SCII for oilseeds} = \frac{120\times5+95\times3}{20\times365+3\times243} = \frac{885}{8029} = 0.11$$

$$\text{SCII for pulses} = \frac{85\times2+75\times2+70\times1}{20\times365+3\times243} = \frac{390}{8029} = 0.05$$

SCII for vegetables

$$= \frac{100\times1+70\times2+120\times2+70\times2+50\times2+120\times1+70\times1+120\times3}{20\times365+3\times243} = \frac{1270}{8029} = 0.16$$

8.8. RELATIVE CROPPING INTENSITY INDEX (RCII)

It is a modification of CII and determines the of area-time allotted to one crop or group of crops relative to the area- time actually used in the production of all the crops.

$$RCII = \frac{\sum_{k=1}^{NK} a_k t_k}{A_0 T + \sum_{i=1}^{m} A_i T_i}$$

Where,

k = 1,2,3,..........................N_k

j = 1,2,3,..........................m

N_k = total number of crops within a specific designation such as vegetable or field crops grown by the farmers during the time period T m   = total number of crop grown by the farmers during the time period T.

a_k = area occupied by kth crop

t_k = duration of the kth crop

a_i = area occupied by ith crop

tj = duration of the ith crop

Example 9. Caculate the RCII of cereals, oilseeds, pulses and vegetables of Examples 6.

Solution:

RCII for cereals

$$= \frac{120 \times 5 + 120 \times 5 + 130 \times 5 + 120 \times 4 + 110 \times 4 + 120 \times 1}{\begin{array}{c} 600 + 600 + 600 + 650 + 480 + 440 + 120 + 110 + 170 + 140 \\ + 240 + 150 + 140 + 100 + 120 + 70 + 70 + [3 \times 120 + 3 \times 95] \end{array}} = \frac{2890}{5445} = 0.53$$

RCII for oilseeds

$$= \frac{120 \times 5 + 95 \times 3}{\begin{array}{c} 600 + 600 + 600 + 650 + 480 + 440 + 120 + 110 + 170 + 140 \\ + 240 + 150 + 140 + 100 + 120 + 70 + 70 + [3 \times 120 + 3 \times 95] \end{array}} = \frac{885}{5445} = 0.16$$

RCII for pulses

$$= \frac{85 \times 2 + 75 \times 2 + 70 \times 1}{\begin{array}{c} 600 + 600 + 600 + 650 + 480 + 440 + 120 + 110 + 170 + 140 \\ + 240 + 150 + 140 + 100 + 120 + 70 + 70 + [3 \times 120 + 3 \times 95] \end{array}} = \frac{390}{5445} = 0.07$$

SCII for vegetables

$$= \frac{100 \times 1 + 70 \times 2 + 75 \times 2 + 120 \times 2 + 70 \times 2 + 50 \times 2 + 120 \times 1 + 70 \times 1 + 120 \times 3}{\begin{array}{c} 600 + 600 + 600 + 650 + 480 + 440 + 120 + 110 + 170 + 140 \\ + 240 + 150 + 140 + 100 + 120 + 70 + 70 + [3 \times 120 + 3 \times 95] \end{array}} = \frac{1270}{5445} = 0.24$$

8.9.DIVERSITY INDEX(DI)

It measures the multiplicity of crops or farm products which are planted in a single year by computing the reciprocals of sum of square of the share of gross revenue received from each individual farm enterprises in a single year.

$$DI = \frac{1}{\sum_{i=1}^{n} a_k t_k \left[y_i \middle/ \sum_{I=1}^{n} a_k t_k \right]}$$

Where,

n = total number of enterprises(crop or farm products)

y_i = gross revenue of the ith enterprises produced within a year

Example 10. A farmer has 5 ha farm with field crops, horticulture, diary, fishery, apiary, piggery, and poultry. His revenue from different enterprises per year is shown below.

Enterprise	Gross revenue(Rs)
Field crops	100000.00
Horticulture	300000.00
Diary	200000.00
Fishery	100000.00
Apiary	20000.00
Piggery	200000.00
Poultry	150000.00

Calculate the Diversity Index (DI) of the farm.

Enterprise	Gross revenue(y_i)	$\left(y_i \Big/ \sum\limits_{l=k}^{n} a_k t_k \right)$	$\left[\left(y_i \Big/ \sum\limits_{l=k}^{n} a_k t_k \right) \right]^2$
Field crops	100000.00	0.0935	0.0087
Horticulture	300000.00	0.2824	0.0786
Diary	200000.00	0.1869	0.0349
Fishery	100000.00	0.0935	0.0087
Apiary	20000.00	0.0187	0.0003
Piggery	200000.00	0.1869	0.0349
Poultry	150000.00	0.1402	0.0197
$\Sigma =$	1070000.00	0.1402	0.1859

$$DI = \frac{1}{0.1859} = 5.38$$

8.10. HARVEST DIVERSITY INDEX (HDI)

It is computed using the same equation as the DI except that the value of each farm enterprises is replaced by value of each harvest

$$HDI = \cfrac{1}{\sum\limits_{i=1}^{n} \left[y_i \Big/ \sum\limits_{l=1}^{n} a_k t_k \right]}$$

Where,

n = total number of crops

Y_i = gross revenue of the ith crop planted and harvested within a year

Example 11. Gross value of the crops planted in example 3 are given on parenthesis after the crop:

Block	Area(ha)	Sequence		
		Summer	Kharif	Rabi
A	5	-	Rice (2,50,000/-)	Mustard (1,80,000/-)
B	5	-	Rice (2,50,000/-)	Rice (3,00,000/-)
C	4	-	Rice (2,00,000/-)	Wheat (2,00,000/-)
D	1	-	Rice (50,000/-)	Tomato (2,00,000/-)
E	2	Greengram (1,20,000/-)	Baby corn (2,50,000/-)	Potato (3,00,000/-)
F	2	Blackgram (1,20,000/-)	Baby corn (2,50,000/-)	Knolkhol (2,00,000/-)
G	1	Okra (1,50,000/-)	Greengram (60,000/-)	Broccoli (3,00,000/-)

Calculate the Harvest Diversity Index.

Solution:

Enterprise	Gross revenue(y_i)	$\left(y_i \Big/ \sum_{l=k}^{n} a_k t_k \right)$	$\left[\left(y_i \Big/ \sum_{l=k}^{n} a_k t_k \right) \right]^2$
Rice	10,50,000	0.3107	0.0965
Mustard	1,80,000	0.0533	0.0028
Wheat	2,00,000	0.0592	0.0035
Tomato	2,00,000	0.0592	0.0035
Greengram	1,80,000	0.0533	0.0028
Baby corn	5,00,000	0.01479	0.0219
Potato	3,00,000	0.0888	0.0079
Blackgram	1,20,000	0.0355	0.0013
Knolkhol	2,00,000	0.0592	0.0035
Okra	1,50,000	0.0444	0.0020
Broccoli	3,00,000	0.0888	0.0079
$\Sigma =$	33,80.000	-	0.1535

$$HDI = \frac{1}{0.1535} = 6.51$$

8.11. SIMULTANEOUS CROPPING INDEX (SCI)

It indicates how sustainable a management practice is. It is computed by-

$$SCI = \frac{HDI \times 10,000}{MCI}$$

Example 12. Calculate the SCI of example 11.

Solution:

　　HDI　　6.51

　　MCI　　225%

$$SCI = \left[6.15 \times \frac{10,000}{225} \right] = 289.3$$

8.12. SUSTAINABILITY INDEX

It indicates how sustainable a management practice is. It is computed by-

$$SYI = \frac{Y - s}{Y_{max}}$$

Where,

　Y　= Average yield of the management practice over years

　σ　= Standard deviation of yield of the management practice over years

Y $_{max}$ = Observed maximum yield of the experiment

Example 13. Calculate the SI of INM practice and 100% ferti9lizer treatment of rice from the following observation:

Solution:

Treatment	Grain yield(q/ha)									
	2001	2002	2003	2004	2005	2006	2007	2008	2009	2010
INM	43.6	47.9	47.3	45.3	46.3	45.6	47.3	46.3	48.3	49.6
100% Fertilizer	45.2	48.3	46.3	42.3	40.2	39.6	38.3	38.9	37.5	37.5

Solution:

Average yield of INM practive over years	= 46.74 q/ha
Standard deviation of yield INP practice	= 1.70q/ha
Average yield of 100% fertilizer practice over years	= 41.41q/ha
Standard deviation of yield of 100% fertilizer practice	= 3.91q/ha
Observed maximum yield of the experiment	= 49.6q/ha

$$\text{SYI of INM} = \frac{46.74-1.70}{49.60} = 0.91$$

$$\text{and SYI of 100\% fertilizer practice} = \frac{41.41-3.91}{49.60} = 0.76$$

Thus, INM practice is more sustainable than 100% fertilizer practice

8.13. SUSTAINABLE VALUE INDEX

It indicates how sustainable a management practice is un term of economic return. It is computed by-

$$\text{SVI} = \frac{N-\sigma}{P_{max}}$$

Where,

$N =$ Average net profit of the management practice over years

$\sigma =$ Standard deviation of net profit of the Management practice over years

$P_{max} =$ Observed maximum net profit of the experiment

Example 14. Calculate the SVI of INM and 100% fertilizer treatment of rice from the following observation:

Treatment	Net profit(Rs/ha)									
	2001	2002	2003	2004	2005	2006	2007	2008	2009	2010
INM	32,320	37,480	36,760	34,360	35,560	34,720	36,760	35,560	37,840	39,520
100% Fertilizer	34,240	37,960	35,560	30,760	28,240	27,520	25,960	26,680	25,000	25,000

Solution:

Average net returnof INM practice over years $\qquad$ = Rs 36,088/ha

Standard deviation of net return of INP practice $\qquad$ = Rs 2,036/ha

Average net return of 100% fertilizer practice over years = Rs 29,692/ha

Standard deviation of net return of 100% fertilizer practice = Rs 4,696/ha

Observed maximum net return of the experiment = Rs 39,520/ha

$$\text{SVI of INM} = \frac{36,088\text{-}2,036}{39,520} = 0.86$$

$$\text{And SVI of 100\% fertilizer practice} = = \frac{29,692\text{-}4,696}{39,520} = 0.63$$

Thus, INM practice is more sustainable than 100% fertilizer practice in terms of economic return, as well.

8.14. CROP EQUIVALENT YIELD(CEY)

The yields of different intercrops or crop sequence are converted into equivalent yield of any crop based on price of the produce. The CEY is estimated as :

$$CEY = \sum_{i=1}^{n} y_i e_i$$

$$e = \frac{P_{bc}}{P_i}$$

where

    y_i = yield of ith component

    e_i = equivalent factor

    P_i = Ptice of the ith crop

    P_{bc} = Price of the crop to which the yield is converted

Example 15. The yield of rice and rapeseed in rice-rapeseed cropping system are 5000kg and 600kg per ha, respectively. If the price of rice and rapeseed are Rs. 10 and Rs. 30 per kg, respectively, convert the yield of the system into (a) Rice Equivalent Yield (b) Rapeseed equivalent Yield

Solution:

(a) Rice equivalent yield of the system

$$= 50,00 \times \frac{10}{10} + 600 \times \frac{30}{10} = 6,800 \text{kg/ha}$$

(b) Rapeseed equivalent yield of the system

$$= 50,00 \times \frac{30}{10} + 600 \times \frac{30}{30} = 2,266.7 \text{kg/ha}$$

Example 16. The yield of rice, potato and greengram in rice-potato-greengram cropping system are 5.4,25.0 and 0.9 tom per ha, respectively. If the price of rice, potato and greengram are Rs. 10,8 and Rs. 60 per kg, respectively, convert the yield of the system into (a) Rice Equivalent Yield (b) potato equivalent Yield and (b) greengram equivalent Yield.

Solution:

(a) Rice equivalent yield of the system

$$= 5,400 \times \frac{10}{10} + 25,000 \times \frac{8}{10} + 900 \times \frac{60}{10}$$

$$= 30,800 \text{ kg/ha} = 30.8 \text{ ton/ha}$$

(b) Potato equivalent yield of the system

$$= 5,400 \times \frac{10}{8} + 25,000 \times \frac{8}{8} + 900 \times \frac{60}{8}$$

$$= 38,500 \text{ kg/ha} = 38.5 \text{ ton/ha}$$

(c) Greengram equivalent yield of the system

$$= 5,400 \times \frac{10}{60} + 25,000 \times \frac{8}{50} + 900 \times \frac{60}{60}$$

$$= 5,133.33 \text{kg/ha} = 5.13 \text{ ton/ha}$$

Example 17. Calculate the wheat equivalent yield of the following intercropping systems:

System	Yield(t/ha)				Price (Rs./ton)			
	Grain	Straw	Grain	Straw	Grain	Straw	Grain	Straw
Wheat + Lentil	3.06	6.94	0.35	0.92	6400	1500	15,250	2500
Wheat + Toria	3.04	6.77	0.32	0.95	6400	1500	16,650	-

Solution:

(a) Wheat equivalent yield of wheat+ lentil intercropping:

$$= 3.06 \times \frac{6400}{6400} + 6.94 \times \frac{1500}{6400} + 0.35 \times \frac{15250}{6400} + 0.92 \times \frac{2500}{6400}$$

$$= 3.06 + 1.63 + 0.83 + 0.36 \quad = 5.88 \text{ ton/ha}$$

(b) Wheat equivalent yield of wheat+ toria intercropping:

$$= 3.04 \times \frac{6400}{6400} + 6.77 \times \frac{1500}{6400} + 0.32 \times \frac{16650}{6400}$$

$$= 3.04 + 1.58 + 0.83 = 5.45 \text{ ton/ha}$$

8.15 RELATIVE YIELD TOTAL (RYT)

The oldest established measure of the yield advantage of crop mixtures is the relative yield total (RYT) index was designed as a measure of the extent to which various crops components shared common resources, rather than as a direct measure of yield advantage. RYT is measured by the expression:

$$\text{Relative Yield Total (RYT)} = = \sum_{i=1}^{m} \frac{Y_{ij}}{Y_{ij}}$$

Where,

$Y_{ij} =$ biomass of ith component from a unit area of intercrop expressed as a fraction of yield

$Y_{ii} =$ biomass of ith component grown as sole crop over the same area

A RYT of 1.0 is said to indicate that the components of the mixture fully share the same limiting resources, i.e. they are fully in competition with each other. Values of RYT = 1.0 would also occur in the total absence of competition, e.g. if the density of the monoculture and mixture were sufficient. Values between 1.o and 2.0 would indicate that the components were only in partial competition with each other. RYT values of less than 1.0 would indicate that the crop components suppressed each other more thn could be accounted for by competition alone, e.g., by allelopathy. A RYT value of 2.0 would be indicate that the components did not share limiting resources at all, i.e., they did not compete at all for limiting resources. RYT values of greater than 2.0 would mean that at least one component actually stimulated the growth of the other.

8.16. LAND EQUIVALENT RATIO(LER)

The most commonly used index of agronomic yield advantage is the land equivalent ratio (LER), this index is in fact identical to RYT, since it is obtained by the expression:

$$\text{LER} = \sum_{i=1}^{m} \frac{Y_{ij}}{Y_{ij}}$$

Y_{ij} = yield of ith component from a unit area of intercrop expressed as a fraction of yield

Y_{ii} = yield of ith component grown as sole crop over the same area

The main difference between LER and RYT is in interpretation, rather than expression, since LER is considered as a measure of the efficiency of grain or economic yield production of the crop mixture, compared with sole crops and based on land use. An LER value of 1.0 indicates that the same amount of land would be required to obtain a given amount of economic yield of each component, regardless of whether the two components were grown in mixture or pure stands. An LER value of 1.2 for example, would indicate that 20 percent more land would be needed to produce a given amount of each of the two crop components in pure stands as in mixtures.

Example 18. Calculate the LER of wheat+ lentil and wheat + toria intercropping system from the following observation:

System	Yield(t/ha)		
	wheat	lentil	toria
Wheat sole	3.33	-	-
Lentil sole	-	1.04	-
Toria sole	-	-	1.12
Wheat+lentil	3.06	0.35	-
Wheat+toria	3.04	-	0.32

Solution:

Wheat+lentil system:

$$\text{LER of wheat} = \frac{3.06}{3.33} = 0.92$$

$$\text{LER of lentil} = \frac{0.35}{1.04} = 0.34$$

LER of the system = 0.92+0.34 = 1.26

Wheat+toria system:

LER of wheat $= \dfrac{3.04}{3.33} = 0.91$

LER of toria $= \dfrac{0.32}{1.12} = 0.29$

LER of the system $= 0.91+0.29 = 1.20$

8.17. PRICE EQUIVALENT RATIO (PER)

Price equivalent ratio(PER) is the ratio of price obtained under intercropping system as compared to the price that could have been obtained under sole cropping. The PER can be mathematically represented as follows:

$$PER = \frac{(Y_{ab} \times Y_{ap}) + (Y_{ba} \times Y_{bp})}{1/2(Y_{aa} \times Y_{ap} + (Y_{bb} \times Y_{bp})}$$

Where,

Y_{aa} = yield of component crop 'a' as sole crop

Y_{bb} = yield of component crop 'b' as sole crop

Y_{ab} = yield of component crop 'a' as intercrop in combination with 'b'

Y_{ba} = yield of component crop 'b' as intercrop in combination with 'a'

Y_{ap} = market price of produce of component crop 'a'

Y_{bp} = market price of produce of component crop 'b'

Example 19. Calculate the PER ofwheat+lentil and wheat+toria intercropping system from the following observation:

System	Yield(t/ha)		
	wheat	lentil	toria
Wheat sole	3.33	-	-
Lentil sole	-	1.04	-

Toria sole	-	-	1.12
Wheat+lentil	3.06	0.35	-
Wheat+toria	3.04	-	0.32
Price(Rs./ton)	64,00	15,250	16,650

Solution:

Wheat+lentil system:

$$PER = \frac{(3.06 \times 6400) + (0.35 \times 15250)}{1/2(3.33 \times 6400 + 1.04 \times 15250)} = 1.34$$

Wheat+toria system:

$$PER = \frac{(3.06 \times 6400) + (0.32 \times 16650)}{1/2(3.33 \times 6400 + 1.12 \times 16650)} = 1.25$$

Example 20. Calculate the PER of wheat+lentil and wheat+toria intercropping system from the following observation:

System	Yield(t/ha)				Price(Rs./ton)			
	Wheat		Lentil/Toria		Wheat		Lentil/Toria	
	Grain	Straw	Grain	Straw	Grain	Straw	Grain	Straw
Wheat sole	3.33	6.58	-		6400	1500	-	-
Lentil sole	-	-	1.04	2.24	-	-	15250	2500
Toria sole	-	-	1.12	2.64	-	-	16650	-
Wheat+lentil	3.06	6.94	0.35	0.92	6400	1500	15250	2500
Wheat+toria	3.04	6.77	0.32	0.95	6400	1500	16650	-

Solution:

Wheat+lentil system:

$$PER = \frac{(3.06 \times 6400) + (6.94 \times 1500) + (0.35 \times 15250) + (0.92 \times 2500)}{1/2(3.33 \times 6400 + 6.58 \times 1500 + 1.04 \times 15250 + 2.24 \times 2500)} = 1.43$$

Wheat+toria system:

$$PER = \frac{(3.04 \times 6400) + (6.77 \times 1500) + (0.32 \times 16650)}{1/2(3.33 \times 6400 + 6.58 \times 1500 + 1.12 \times 16650)} = 1.40$$

8.18. RELATIVE CROWDING COEFFICIENT (RCC)

RCC is used in replacement series experiments. Each component has its own coefficient (K) which gives a measure of whether that component has produced more, or fewer yields than expected. It is given by-

$$K_{ab} = \frac{Y_{ab} \times Z_{ba}}{(Y_{aa} - Y_{ab}) \times Z_{ab}}$$

$$K_{ba} = \frac{Y_{ba} \times Z_{ab}}{(Y_{bb} - Y_{ba}) \times Z_{ba}}$$

$K = K_{ab} \times K_{ba}$

Where,

K_{ab} = RCC of component crop 'a' in the mixture

Y_{bb} = RCC of component crop 'b' in the mixture

K = RCC of the system

Y_{aa} = yield of component crop 'a' as sole crop

Y_{bb} = yield of component crop 'b' as sole crop

Y_{ab} = yield of component crop 'a' as intercrop in combination with 'b'

Y_{ba} = yield of component crop 'b' as intercrop in combination with 'a'

Z_{ba} = sown proportion of component 'b' in combination with 'a'

Z_{ab} = sown proportion of component 'a' in combination with 'b'

If a component has RCC,

>1, it means that it has produced more yield than expected yield

=1, it means that it has produced equal yield than expected yield

<1, it means that it has produced less yield than expected yield

Moreover, the component with higher coefficient is dominant one.

Besides, if K,

>1, there is yield advantage

=1, no difference (no yield advantage)

<1, yield disadvantage

Example 21. Calculate the RCC of wheat+lentil and wheat+toria intercropping system from the following observation, if the crops are sown at 2:1 row ratio of wheat and lentil/toria

System	Yield(t/ha)		
	wheat	**lentil**	**toria**
Wheat sole	3.33	-	-
Lentil sole	-	1.04	-
Toria sole	-	-	1.12
Wheat+lentil	3.06	0.35	-
Wheat+toria	3.04	-	0.32
Price(Rs./ton)	64,00	15,250	16,650

Solution:

In both the sequence, crops are sown at 2:1 row ratio. i.e. 66.67% and 33.333% area are occupied by wheat and lentil/toria, respectively.

Wheat+lentil system:

$$K_{wheat} = \frac{(3.06 \times 0.3333)^*}{(3.33 - 3.06) \times 0.6667} = 5.66$$

$$K_{lentil} = \frac{(0.35 \times 0.6667)^*}{(1.04 - 0.35) \times 0.3333} = 1.01$$

$$K_{system} = K_{wheat} \times K_{lentil} = 5.66 \times 1.01 = 5.72$$

*instead of 0.3333 and 0.6667, if you put 1 and 2 or 33.33 and 66.67, you will get the same result.

[Here, both K_{wheat} and K_{lentil} is greater than 1. This indicated that both the crops had produced higher yield than expected yield. Since, $K_{wheat} > K_{lentil}$, K_{wheat} is dominant component. K value greater than 1 of the system, indicates yield advantage of the system.]

Wheat+toria system:

$$K_{wheat} = \frac{(3.04 \times 0.3333)}{(3.33 - 3.04) \times 0.6667} = 5.24$$

$$K_{toria} = \frac{(0.32 \times 0.6667)}{(1.12 - 0.32) \times 0.3333} = 0.80$$

$$K_{system} = K_{wheat} \times K_{toria} = 5.24 \times 0.80 = 4.19$$

[Here, $K_{wheat} > 1$ and $K_{lentil} < 1$. This indicated wheat had produced higher yield than expected yield while toria had produced less than the expected yield. Since $K_{wheat} > K_{toria}$, K_{wheat} is dominant component. K value greater than 1 of the system indicates yield advantage of the system.]

8.19. AGGRESSIVITY (A)

Aggressivity shows the relationship between dominant and dominated species grown together, especially when any of the two offers competition. It may be calculated as-

$$\frac{\text{Mixture yield of a}}{\text{Expected yield of a}} - \frac{\text{Mixture yield of b}}{\text{Expected yield of b}}$$

$$A_{ab} = \frac{Y_{ab}}{Y_{ab} \times Z_{ab}} - \frac{Y_{ba}}{Y_{bb} \times Z_{ba}}$$

Where,

Y_{ab} = Agressivity of 'a' in the mixture over 'b'

Y_{aa} = yield of component crop 'a' as sole crop

Y_{bb} = yield of component crop 'b' as sole crop

Y_{ab} = yield of component crop 'a' as intercrop in combination with 'b'

Y_{ba} = yield of component crop 'b' as intercrop in combination with 'a'

Z_{ba} = sown proportion of component 'b' in combination with 'a'

Z_{ab} = sown proportion of component 'a' in combination with 'b'

If Agressivity,

=0, component crops are equal competitive

>0, component is dominant

<1, component is dominated by the other

Moreover, greater the numerical value, bigger is the difference in competitive abilities and bigger the difference between actual and expected yields.

Example 22. Calculate the aggressivity of wheat+lentil and wheat+toria intercropping system on the basis of the information given in example 17.

Solution:

In both the sequence, crops are sown at 2:1 row ratio. i.e. 66.67% and 33.333% area are occupied by wheat and lentil/toria, respectively.

Wheat+lentil system:

$$A_{wheat} = \frac{3.06}{3.33 \times 0.6667} - \frac{0.35}{1.04 \times 0.3333}$$

$$= 1.38 - 1.01 = 0.37$$

$$A_{lentil} = \frac{0.35}{1.04 \times 0.3333} + \frac{3.06}{3.33 \times 0.6667}$$

$$= -1.01 + 1.38 = -0.37$$

[Here, Aggressivity of wheat is +ve, which means that wheat is the dominant component in the mixture. Aggressivity of Lentil is −ve. Thus, lentil is dominated component.

Again, if we put 1 and 2 in place of 0.3333 and 0.6667, the result will be $1/3^{rd}$ of the Aggressivity value tht we have obtained. i.e. A_{wheat} and A_{toria} will be +0.12 and -0.12 respectively. Nothing wrong, we need to know whether A is +ve or −ve. Moreover, absolute value will decrease or increase proportionately for all the cropping systems compared. Thus, it will not misinterpret the difference in competitive abilities. However, in most of the cases people prefer the former ones.]

Wheat+lentil system:

$$A_{wheat} = \frac{3.04}{3.33 \times 0.6667} - \frac{0.32}{1.12 \times 0.3333}$$

$$= 1.37 - 0.86 \quad = 0.51$$

$$A_{toria} = \frac{0.32}{1.12 \times 0.3333} + \frac{3.04}{3.33 \times 0.6667}$$

$$= -0.86 - 1.37 \quad = -0.51$$

[Here, Aggressivity of wheat is +ve, which means that wheat is the dominant species in the mixture. Aggressivity of toria is −ve. Thus, toria is dominated species.]

8.20. COMPETITION RATIO(CR)

Competition ratio can be written as-

$$CR_a = \frac{Y_{ab}}{Y_{ab} \times Z_{ab}} \div \frac{Y_{ba}}{Y_{bb} \times Z_{ba}}$$

$$= \frac{LER_a}{LER_b} \times \frac{Z_{ba}}{Z_{ab}}$$

Where,

CR_a = Competition Ratio of 'a' in the mixture over 'b'

Y_{aa} = yield of component crop 'a' as sole crop

Y_{bb} = yield of component crop 'b' as sole crop

Y_{ab} = yield of component crop 'a' as intercrop in combination with 'b'

Y_{ba} = yield of component crop 'b' as intercrop in combination with 'a'

Z_{ba} = sown proportion of component 'b' in combination with 'a'

Z_{ab} = sown proportion of component 'a' in combination with 'b'

LER_a = LER of component 'a'

LER_b = LER of component 'b'

Example 23. Calculate the Competition Ratio of wheat+lentil intercropping system on the basis of the information given in example 17.

Solution:

Wheat + lentil system:

$$CR_{wheat} = \frac{3.06}{3.33 \times 2} \div \frac{0.35}{1.04 \times 1}$$

$$= \frac{0.4594}{0.3365} = 1.37$$

$$CR_{lentil} = \frac{0.35}{1.04 \times 1} \div \frac{3.06}{3.33 \times 2}$$

$$= \frac{0.4594}{0.3365} = 0.73$$

[Here, CR_{wheat} is 1.37. this means that wheat produced 1.37 times as much as expected yield and is 1.37 times as competitive.]

⋮EXERCISES ▶

1. Calculate the rotational intensity of the following crop sequences:

   a) Rice- Wheat   (1 year)

   b) Rice- Wheat-Cowpea   (1 year)

   c) Sesamum- Greengram-Broccoli-Blackgram-
      Greengram-Cabbage   (2 year)

   d) Maize-Potato-Sugarcane-Ratoon-Blackgram   (3 year)

   e) Maize-Sugarcane+ Potato-Ratoon+Mustard-
      Blackgram   (3 year)

2. Calculate the MCI of each block and cropping intensity of a 14 ha farm where land is utilized as follows:

Block	Area(ha)	Sequence		
		Summer	Kharif	Rabi
A	4	-	Rice	Mustard
B	4	-	Rice	Rice
C	3	-	Rice	Wheat
D	0.5	-	Rice	Tomato
E	1	Greengram	Baby corn	Potato
F	1	Blackgram	Baby corn	Knolkhol
G	0.5	Okra	Greengram	Broccoli

3. A farmer has 16 ha cultivable land available for entire year. In addition, he has 3 ha flood prone Char land which is temporarily available to him for cultivation from October to May(243 days). The land utilized as follow:

Block	Area(ha)	Sequence		
		Summer	Kharif	Rabi
A	4	-	Rice(120)*	Mustard(120)
B	4	-	Rice(120)	Rice(130)

C	3	-	Rice(120)	Wheat(110)
D	0.5	-	Rice(120)	Tomato(110)
E	1	Greengram(85)	Baby corn(70)	Potato(120)
F	1	Blackgram(75)	Baby corn(70)	Knolkhol(50)
G	0.5	Okra(120)	Greengram(70)	Broccoli(70)
Char land	2	Vegetable(120)	-	Rapeseed(95)

*Duration occupied by the crop in the field

Calculate

a. LUE of each block

b. CII of the farm

c. SCII of cereals (rice and wheat), oilseeds (rapeseed and mustard), pulses (greengram and blackgram) and vegetables.

d. RCII of cereals (rice and wheat), oilseeds (rapeseed and mustard), pulses (greengram and blackgram) and vegetables.

4. A farmer has 5 ha farm with field crops, horticulture, diary, fishery, apiary, piggery, and poultry. His revenue from different enterprises per year is shown below.

Enterprise	Gross revenue(Rs)
Field crops	120000.00
Horticulture	350000.00
Diary	300000.00
Fishery	110000.00
Apiary	25000.00
Piggery	230000.00
Poultry	170000.00

Calculate the Diversity Index (DI) of the farm.

5. Gross value of the crops planted in a 12 ha are given on parenthesis after the crop in Rs.:

Block	Area(ha)	Sequence		
		Summer	Kharif	Rabi
A	5	-	Rice(2,00,000/-)	Mustard(1,50,000/-)
B	5	-	Rice(2,00,000/-)	Rice(2,50,000/-)
C	4	-	Rice(2,60,000/-)	Wheat(1,60,000/-)
D	1	-	Rice(45,000/-)	Tomato(80,000/-)

Calculate

    a.   Harvest Diversity Index of the farm

    b.   Multiple cropping intensity(cropping intensity) of the farm

    c.   SCI of the farm

6. The yield and value of kharif rice with rice –fallow and rice-rice sequence are given below

	2008		2009		2010		2011		2012	
Seque-nce	Yield (t/ha)	Value (Rs.)	Yield (t/ha)	Value (Rs.)	Yield (t/ha)	Value (Rs.)	Yield (t/ha)	Value (Rs.)	Yield (t/ha)	Value (Rs.)
rice – fallow	4.8	48,000	4.7	47,000	4.7	47,000	4.8	48,000	4.6	46,000
rice-rice	4.8	48,000	4.5	45,000	4.3	43,000	4.1	41,000	3.9	39,000

Calculate

    (a) Sustainable Index(SI) of rice –fallow and rice-rice cropping system

    (b) Sustainable Value Index (SVI) of rice –fallow and rice-rice cropping system, if cost of cultivation of rice is Rs.25,000/ha.

7. The yield of rice and rapeseed in rice-rapeseed cropping system are 48,000 kg and 630 kgper hectare, respectively. If the price of rice and rapeseed are Rs. 11 and 30 per kg, respectively, convert the yield of the system into **(a)** Rice equivalent yield   **(b)** Rapeseed equivalent yield

8. The yield of rice,potato and greengram in rice-potato-greengram cropping system are 4.93, 24.15 and 0.85 ton per hectare, respectively.

If the price of rice, potato and greengram are Rs. 11, 9 and 70, respectively, convert the yield of the system into **(a)** Rice equivalent yield **(b)** Potato equivalent yield and **(c)** Greengram equivalent yield.

9. Yield and price of different products of wheat + Lentil and Wheat + toria intercropping system sown at 2:1 ratio of wheat and lentil/toria are given below:

System	Yield(t/ha)				Price(Rs./ton)			
	Wheat		Lentil/Toria		Wheat		Lentil/Toria	
	Grain	Straw	Grain	Straw	Grain	Straw	Grain	Straw
Wheat sole	3.50	6.81	-	-	6,500	1,000	-	-
Lentil sole	-	-	1.10	2.33	-	-	50,000	2,000
Toria sole	-	-	1.18	2.75	-	-	30,000	-
Wheat+lentil	3.10	7.10	0.36	0.95	6,500	1,000	50,000	2,000
Wheat+toria	3.00	6.70	0.33	0.97	6,500	1,000	30,000	-

Calculate

(a) Wheat equivalent yield

(b) LER

(c) PER

(d) RCC

(e) Aggressivity

(f) Competition Ratio of wheat + Lentil and Wheat + toria intercropping system.

⋮ ANSWERS ▶

1. (a) 200% (b) 300% (c) 250% (d) 167% (e) 133%
2. Block A=200% Block B=200% Block C= 200%
   Block D =200%
   Block E=75.34 % Block F=53.42% Block G=300%
   Whole Farm = 217.86%

3. **(a)** Block A=65.75% Block B=68.49%

    Block C=63.01 % Block D= 60.27%

    Block E= 75.34% Block F=53.42%

    Block G=71.23% Char area= 58.90%

3. **(b)** CII=0.68

3. **(c)** Cereals=0.40;Oilseeds=0.12;Pulses=0.04;Vegetables=0.13

3. **(d)** Cereals=0.59;Oilseeds=0.18;Pulses=0.05;Vegetables=0.18

4. 5.30

5. **(a)** 1.97 **5.** **(b)** 225%   **5.** **(b)** 86.7%

6. **(a)** SYI of rice-fallow=0.97 and rice-rice= 0.83

6. **(a)** SVI of rice-fallow=0.93 and rice-rice= 0.64

CHAPTER

9

WATER MANAGEMENT

Water is a very precious resource. No living being can survive without water. About 69% of the total water used in the world is spent in agriculture and allied activities. A major part of it is used for irrigation. However, due to increasing population and multifarious use, per capita availability and shared of water for irrigation is decreasing day by day. This necessitates the precise application of irrigation water to get maximum benefit of applied water.

9.1. VOLUME AND MASS RELATIONSHIP OF SOIL

Soil is a three phase system – solids, water and air. The volume and mass relationships among these three phases are useful for characterizing the physical condition of the soil.

9.1.1. Total Volume of Soil (V_t)

$$V_t = V_s + V_v$$
$$= V_s + V_a + V_w$$

Where,

V_t = Total volume of soil

V_s = Volume of solid

V_v = volume of pore (void) = $V_a + V_w$

V_a = Volume of air

V_w = Volume of water

9.1.2. Total Mass of Soil (M_t)

$$M_t = M_s + M_v$$

Where,

M_t = Total mass of soil or mass of wet soil

M_s = Mass of soil solid

M_w = Mass of soil water

[mass of air is considered as negligible]

Example 1. If the volume of soil solid is 58 cc, and total volume of soil is 110 cc, calculate the volume of pore (void).

Solution :

Volume of pore (void) = (110 – 58) cc

= 52 cc

Example 2. If the volume of water in Example 1 is 25 cc, calculate the volume of air.

Solution :

Volume of air = (52 – 25) cc

= 27 cc

9.1.3. Void Ratio (e)

$$e = \frac{V_v}{V_z}$$

Example 3. Calculate void ratio of Example 1.

Solution :

Given that, V_v = 52 cc

V_s = 58 cc

Void ratio (*e*) $= \dfrac{52}{58}$

= 0.90

9.1.4. Porosity (*f*)

The pore space or porosity is that proportion of the total soil volume which is occupied by air and/or water. It is generally expressed in percentage. It may also be expressed as the total volume of pores per unit volume of soil (cc/cc).

$$f = \frac{V_v}{V_t}$$

Where

V_v = Volume of pore (void)

V_t = Volume of soil

Example 4. Calculate porosity of Example 1.

Solution :

Given that, V_v = 52 cc

V_t = 110 cc

Total porosity (*f*) $= \dfrac{52}{110}$

= 0.4727 cc/cc

or Total porosity = 47.27 %

9.1.5. Gravimetric Moisture Content (·§)

$$= \frac{M_w}{M_s} \times 100 \%$$

Where,

M_w = Mass of water

M_s = Mass of solid

[Remember, gravimetric moisture content is always determined on the basis of dry weight of soil, *i.e.* soil mass; not on the basis of total soil mass (mass of soil solid + water)]

Example 5. Find out the gravimetric moisture percent of soil from the following data:

Mass of soil = 61 g

Mass of water = 12.1 g

Solution:

Gravimetric moisture content, $= \left[12.1 \times \dfrac{100}{61} \right] \%$

$= 19.84 \%$

Example 6. Find out the gravimetric moisture percent of soil from the following data:

Mass of moisture box	= 39.1 g
Moisture of wet soil with moisture box	= 112.2 g
Mass of oven dry soil with moisture box	= 100.1 g

Solution:

Mass of oven dry soil, M_s = 100.1 − 39.1

$= 61.0$ g

Mass of wet soil   = 112.2 − 39.1

$= 73.1$ g

Mass of water, M_w = 73.1 – 61.0

                                                  = 12.1 g

Gravimetric moisture content, $= \left[12.1 \times \dfrac{100}{40} \right] \%$

                                                  = 19.84 %

9.1.6. Volumetric Moisture Content (Θ)

$$\Theta = \frac{V_w}{V_t} \times 100 \%$$

Where,

V_w = volume of water

V_t = total volume of soil

Example 7. Find out the volumetric moisture percent of soil from the following data:

Volume of soil   = 40 cc

Volume of water = 12.1 cc

Solution:

Volumetric moisture content, $\Theta = \left[12.1 \times \dfrac{100}{40} \right] \%$

$$= 30.25 \%$$

Example 8. Find out the volumetric moisture percent of soil from the following data:

Mass of moisture box   = 39.1 g

Moisture of wet soil with moisture box   = 112.2 g

Mass of oven dry soil with moisture box   = 100.1 g

Volume of soil   = 40 cc

Solution:

Mass of oven dry soil, M_s = 100.1 – 39.1

= 61.0 g

Mass of wet soil, M_t = 112.2 – 39.1

= 73.1 g

Mass of water, M_w = 73.1 – 61.0

= 12.1 g

Volume of water, V_w = 121.1 cc

Volume of soil, V_s = 40.0 cc

Volumetric moisture content, Θ = $\left[12.1 \times \dfrac{100}{40} \right]$ %

= 30.25 %

9.1.7. Degree of Saturation (*s*)

The degree of saturation is the volume of pores occupied by water. Due to entrapped air, complete saturation (*i.e.* 100%) is seldom attained.

$$s = \frac{V_w}{V_v} \times 100 \ \%$$

Example 9. Calculate the degree of saturation of Example 2.

Solution:

Volume of water, V_w = 25

Volume of void, V_v = 52

Degree of saturation, $s = \dfrac{25}{52} \times 100 \ \%$

= 48.08 %

9.1.8. Bulk Density (ρ_b)

Bulk density or dry bulk density is the mass of dry soil per unit volume of soil and is expressed in g/cc or Mg/m^3 (megagram per cubic metre). This bulk volume of soil is the sum of the volume occupied by the soil solids and the volume occupied by voids.

Bulk density is sometimes expressed as wet bulk density. It is the mass of field soil sample per unit bulk volume of the soil including the air space. The mass of field soil sample is the sum of mass of soil solids and mass of water.

If the prefix 'dry' and 'wet' is not mentioned, it is understood that dry bulk density is referred.

$$\text{Dry } \rho_b \text{ (or } \rho_b) = \frac{M_s}{V_t}$$

$$\text{Wet } \rho_b = \frac{M_s + M_s}{V_t}$$

Example 10. If the volume of the soil sample of Ex. 5 is 40 cc, calculate the dry bulk density of the soil.

Solution:

$$\text{Bulk density, } \rho_b = \frac{61}{40}$$

$$= 1.525 \text{ g/cc}$$

Example 11. Find out the dry bulk density of soil from the following data:

(a) Mass of moisture box   = 40 g

(b) Mass of oven dry soil with moisture box = 440 g

(c) Length of core sampler   = 10 cm

(d) Diameter of the core sampler   = 6 cm

Solution:

Mass of oven dry soil, $\quad M_s = (440 - 40)$ g

$= 400$ g

Length of the core sampler, $h = 10$ cm

Diameter of core sampler, $d = 6$ cm

Radius of core sampler, $r = 3$ cm

Volume of soil, $\quad V_t = <!r^2h$

$= 3.1416 \times 3^2 \times 10$

$= 282.74$

Bulk density, $\qquad \rho_b = \dfrac{400}{282.74}$

$= 1.41$ g/cc

9.1.9. Apparent Specific Gravity (ASG)

$$ASG = \frac{\rho_b}{\rho_w}$$

Where,

ρ_b = Bulk density of soil

ρ_w = Density of water

Relationship between volumetric moisture content with bulk density and apparent specific gravity:

$$\theta = \frac{V_w}{V_v} \times 100$$

$$= \frac{M_w}{\rho_w V_t} \times 100 \text{ [where, } \rho_w = \text{density of water]}$$

$$= \frac{M_w M_s}{\rho_w V_t M_s} \times 100 \text{ [Dividing numerator and denominator by } M_s]$$

$$= \left[\frac{M_w}{M_s} \times 100\right] \times \frac{M_s}{\rho_w V_t}$$

$$= \cdot\S \times \times \frac{\rho_b}{\rho_w} \quad \text{[where, } \rho_b = \text{bulk density]}$$

$$= \cdot\S \times ASG \quad \text{[where, ASG = Apparent specific gravity]}$$

Volumetric moisture content = Gravimetric moisture content × Apparent specific gravity

Since, Re value of ρ_w is approximately 1, θ can be obtained by multiplying $\cdot\S$ by the value of ρ_b

Volumetric moisture content = Gravimetric moisture content × Value of bulk density gravity

Many students convert gravimetric moisture content to volumetric moisture content by multiplying gravimetric moisture content by bulk density. Is it a right procedure ? Let me explain.

Say, gravimetric moisture content is 20% and bulk density is 1.5 g/cc. if 20% is multiplied with 1.5 g/cc, what should we get – 30% or 30 g/cc ? Since the Apparent specific gravity does not have any unit, if 20% is multiplied with 1.5, it will give 30%.

Thus, we should not write :
Volumetric moisture content = Gravimetric moisture content × bulk density
We should write :
Volumetric moisture content = Gravimetric moisture content × value of bulk density
(when expressed in g/cc or Mg/m³

> *OR*
>
> *Volumetric moisture content = Gravimetric moisture content ×*
> *Apparent specific gravity*

Example 12. If the gravitational moisture content and bulk density of soil is 19.84 % and 1.525 g/cc, calculate the volumetric moisture content.

Solution :

Given that, gravitational moisture content of the soil = 19.84 %

Bulk density of the soil = 1.525 g/cc

+. Apparent specific gravity of the soil = 19.84 × 1.525 %

= 30.26 %

9.1.10. Particle Density (ρ_s)

Particle density is the mass of dry soil (soil solid) per unit volume of soil solid and is expressed in g/cc or Mg/m^3 (megagram per cubic metre). The particle density of minral soil with small percentage of organic matter (Â1 %) is about 2.65 g/cc.

$$\rho_s = \frac{M_s}{V_s}$$

The actual particle density of soil (ρ_s) having high amount of organic matter is the sum of the density of minerals (ρ_m) multiplied by fraction of minerals (f_m) and density of organic matter (ρ_o) multiplied by fraction of organic matter (f_o).

$\rho_s = \rho_m f_m + \rho_o f_o$

$= (\rho_{m1} f_{m1} + \rho_{m2} f_{m2} + \rho_{m3} f_{m3} + \ldots\ldots + \rho_{mx} f_{mx}) + \rho_o f_o$

Where, $\rho_{m1}, \rho_{m2}, \ldots\ldots \rho_{mx}$ and $f_{m1}, f_{m2}, \ldots\ldots\ldots f_{mx}$ are respective densities of soil constituents 1,2, ... x.

Example 13. If the mass of soil particles and their volume are 111.5 g and 42.1 cc, calculate the particle density of the soil.

Solution :

Mass of soil solid, M_s = 111.5 g

Volume of soil solid, M_s = 42.1 cc

+.   Particle density $= \dfrac{111.5}{42.1}$ g/cc

$= 2.65$ g/cc

Example 14. If the particle density of soil minerals and density of organic matter of soil having organic matter content 0.65 % are 2.65 g/cc and, 1.30 g/cc, calculate the particle density of the soil.

Solution :

Given that, organic matter content $= 0.65$ %

+.   Mineral matter content $= (100 - 0.65)\%$

$= 99.35$ %

+.   Particle density $= \left[2.65 \times \dfrac{99.35}{100}\right] + \left[1.30 \times \dfrac{0.65}{100}\right]$

$= 2.632 + 0.008$

$= 2.64$ g/cc

Example 15. Calculate the particle density of soil having following composition :

Soil constituents	Content	Particle density (g/cc)
Humus	0.58	1.30
Quartz	57.42	2.65
Biotite	42.00	2.95

Solution :

$$\rho_s = \rho_{m1} f_{m1} + \rho_{m2} f_{m2} + \rho_o f_o$$

$$= 2.65 \times \dfrac{57.42}{100} + 2.95 \times \dfrac{42}{100} + 1.3 \times \dfrac{0.58}{100}$$

$$= 1.5216 + 1.239 + 0.0075$$

$$\approx 2.77 \ \text{g/cc}$$

Example 16. Calculate the mass of dry soil of 1 ha furrow slice (15 cm) having bulk density 1.5 g/cc.

Solution :

$$\rho_b = \frac{M_s}{V_t}$$

or $\qquad M_s = \rho_b \times V_t$

Given that, $\quad \rho_b = 1.5 \ \text{g/cc}$

$$= 1.5 \ \text{Mg/m}^3$$

$$V_t = 10{,}000 \ \text{m} \times 0.15 \ \text{m}$$

$$= 1{,}500 \ \text{m}^3$$

$$M_s = 1.5 \ \text{Mg/m}^3 \times 1{,}500 \ \text{m}^3$$

$$= 2{,}250 \ \text{Mg}$$

$$= 22{,}50{,}000 \ \text{kg}$$

$$= 2.25 \times 10^6 \ \text{kg}$$

Relationship between various parameters

1. Relationship between bulk density (ρ_b), particle density (ρ_s) and porosity (f) :

$$f = 1 - \frac{\rho_b}{\rho_s}$$

2. Relationship between porosity (f) and void ratio (e) :

$$f = \frac{e}{1+e}$$

3. Relationship between volumetric moisture (Θ) and gravimetric moisture content ($\cdot\S$) :

$$\theta = \frac{\rho_b}{\rho_b}$$

4. Relationship between volumetric moisture (Θ), porosity (f) and degree of saturation (s) :

$$\theta = f.s$$

5. Relationship between volumetric moisture content (), fraction of air content (f) and porosity (f) :

$$f_a = f - \Theta$$

Example 17. Calculate the total porosity of soil, if bulk density and particle density of soil are 1.35 g/cc and 2.65 g/cc, respectively.

Solution:

$$\text{Total porosity,} \quad f = 1 - \frac{1.35}{2.35}$$

$$= 0.4906 \text{ cc/cc}$$

$$= 49.06 \%$$

Example 18. Calculate the total porosity of soil if void ratio is 0.963.

Solution:

$$f = \frac{e}{1+e}$$

$$f = \frac{0.963}{1+0.963} \text{ cc/cc}$$

$$= 0.4906 \text{ cc/cc}$$

$$= 49.06 \%$$

Example 19. Calculate the volumetric moisture content, if gravimetric moisture content and bulk density of soil is 20% and 1.5 g/cc.

Solution :

Gravimetric moisture content, $\Theta = (20 \times 1.5)$ %

$$= 30 \text{ %}$$

Example 20. Calculate the degree of saturation, if volumetric moisture content and total porosity of soil is 30% and 49.06 %.

Solution :

*.

$$\theta = f.s$$

+.

$$s = \frac{\theta}{f}$$

$$= 30 \times \frac{100}{49.06} \text{ %}$$

$$= 61.15 \text{ %}$$

Example 21. Calculate the fraction of air content (f_a), if volumetric moisture content (Θ) and total porosity (f) are 30% and 49.06 %.

Solution :

Fraction of air, $f_a = f - \Theta$

$$= (49.06 - 30) \text{ %}$$

$$= 19.06 \text{ %}$$

Example 22. Calculate the gravimetric and volumetric moisture content and degree of saturation from the following information:

(a) Mass of wet soil   = 240 g
(b) Mass of dry soil   = 200 g
(c) Bulk density   = 1.35 g/cc
(d) Particle density   = 2.65 g/cc

Solution :

Mass of dry soil, $M_s = 200$ g

Mass of wet soil, $M_t = 240$ g

Mass of water, $M_w = (240 - 200)$ g $= 40$ g

Bulk density $\dfrac{M_s}{M_t} = 1.35$

or $\dfrac{200}{V_t} = 1.35$

or $V_t = \dfrac{200}{1.35}$

$= 148.15$ cc

Total porosity, $f = 1 - \dfrac{\rho_b}{\rho_s}$

$= 1 - \dfrac{1.35}{2.65}$

$= 0.4906$ cc/cc

Thus $\dfrac{V_v}{V_t} = 0.4906$ cc/cc

$V_v = 0.4906 \times V_t$

$= 0.4906 \times 148.15$

$= 72.68$ cc

Mass of water, $M_s = (240 - 200)$ g $= 40$ g

+. Volume of water, $V_w = 40$ cc

and degree of saturation, $s = \dfrac{V_w}{V_v} \times 100$ %

$$= \frac{40}{72.68} \times 100 \%$$

$$= 55.04 \%$$

Gravimetric moisture content $= \dfrac{M_s}{M_t}$

$$= \frac{40}{200} \times 100 \%$$

$$= 20 \%$$

Volumetric moisture content $= \dfrac{V_s}{V_t}$

$$= \frac{40}{148.15} \times 100 \%$$

$$= 27 \%$$

Example 23. Calculate the gravimetric and volumetric moisture content, bulk density, void ratio, porosity, degree of saturation and air filled porosity from the following information:

(a) Mass of wet soil   = 240 g
(b) Mass of dry soil   = 200 g
(c) Volume of soil   = 148 g/cc
(d) Particle density   = 2.65 g/cc

Solution :

Given that, M_t = 240 g
                   M_s = 200 g
                   V_t = 148 cc
                   ρ_s = 2.65 g/cc

+.   Mass of water, M_w = (240 – 200) g = 40 g
     Volume of water, V_w = 40 cc

+. Gravimetric moisture content, $\cdot\S = \dfrac{40}{200} \times 100\ \%$

$$= 20\ \%$$

Volumetric moisture content $\quad \Theta = \dfrac{40}{148} \times 100\ \%$

$$= 27.03\ \%$$

Bulk density of soil, $\quad \rho_b = \dfrac{200}{148}\ \text{g/cc}$

$$= 1.35\ \text{g/cc}$$

Again, $\quad \rho_s = \dfrac{M_s}{V_s}$

or $\quad 2.65 = \dfrac{200}{V_s}$

or $\quad V_s = \dfrac{200}{2.65}$

$$= 75.47\ \text{cc}$$

$$V_v = V_t - V_s$$
$$= 148 - 75.47$$
$$= 72.53\ \text{cc}$$

+. Void ratio, $\quad e = \dfrac{V_v}{V_s}$

$$= \dfrac{72.53}{75.47}$$
$$= 0.96$$

Porosity, $\quad f = \dfrac{V_v}{V_t}$

$$= \frac{72.53}{148} \times 100\ \%$$

$$= 49.01\ \%$$

Degree of saturation $\qquad s \quad = \dfrac{V_w}{V_v} \times 100\ \%$

$$= \frac{40}{72.53} \times 100\ \%$$

$$= 55.15\ \%$$

Air filled porosity, $\qquad f_a \quad = f - \Theta$

$$= (49.01 - 27.03)\ \%$$

$$= 21.98\ \%$$

Example 24. Calculate the bulk density of soil from the following information:

(a) Mass of dry soil $\qquad\qquad\qquad$ = 200 g

(b) Mass of saturated soil $\qquad\quad$ = 272.6 g

(c) Particle density $\qquad\qquad\quad$ = 2.65 g/cc

Solution :

Given that, $\qquad\qquad\qquad\quad$ M_s = 200 g

$\qquad\qquad\qquad\qquad\qquad\quad$ M_t = 272.6 g

$\qquad\qquad\qquad\qquad\qquad\quad$ ρ_s = 2.65 g/cc

+.   Mass of water, $\qquad\qquad$ M_w = (272.6 − 200) g

$\qquad\qquad\qquad\qquad\qquad\qquad$ = 72.6 g

Volume of water, $\qquad\qquad$ V_w = 72.6 cc

*.   Soil is saturated, hence, volume of pore space,

$\qquad\qquad\qquad\qquad\qquad\quad$ V_v = 72.6 cc

Again, particle density, $\qquad$ $\rho_s \quad = \dfrac{M_s}{V_s}$

or $\qquad\qquad\qquad\qquad\qquad$ $2.65 \quad = \dfrac{200}{V_s}$

or $\qquad$ $V_s = 75.47$ cc

+. Total volume of soil, $\qquad V_t = V_s + V_w$

$$= (75.47 + 72.6) \text{ cc}$$

$$= 148.07 \text{ cc}$$

+. Bulk density of soil, $\qquad \rho_b = \dfrac{M_s}{V_s}$

$$= \dfrac{200}{148.07}$$

$$= 1.35 \text{ g/cc}$$

9.2. CALCULATION OF IRRIGATION WATER

Although we calculate the moisture content of soil in percentage, it has to be converted to depth, as irrigation recommendations are generally given in terms of depth, e.g. 4 cm irrigation, 5 cm irrigation, 6 cm irrigation etc. When irrigation water is applied in a particular crop field, it is finally converted to volume.

Conversion of moisture content to depth

If a soil contains 30% volumetric moisture, it means that 100 cc soil contains 30 cc water.

$$i.e. \text{ moisture content} = \dfrac{30 \text{ cc}}{100 \text{ cc}}$$

$$= \dfrac{30 \text{ cm.cm.cm}}{100 \text{ cm.cm.cm}}$$

$$= \dfrac{30 \text{ cm}}{100 \text{ cm}}$$

Thus, 100 cm deep soil contains $\qquad$ = 30 cm water

Hence, 1 cm deep soil contains $\qquad = \dfrac{30}{100}$ cm water

Hence, D cm deep soil contains $\qquad = D \times \dfrac{30}{100}$ cm water

---

Thus, Amount of water content (cm) = Volumetric moisture content (%)
× Soil depth (cm)

Example 25. Calculate the amount of water content in cm, in 0-30 cm soil profile, if the soil contains 25 % moisture by volume.

Solution :

$$\text{Amount of water content} = 25 \times \frac{30}{100} \text{ cm}$$

$$= 7.5 \text{ cm}$$

Example 26. Calculate the amount of water content in cm, in 0-20 cm soil profile, if the soil with bulk density 1.35 g/cc contains 20 % gravimetric moisture.

Solution :

Given that, gravimetric moisture content,

$$\cdot\S \; = 20 \text{ \%}$$

Bulk density, $\qquad\qquad\qquad \rho_b = 1.35$ g/cc

+. Apparent specific gravity $\qquad = 1.35$

+. Volumetric moisture content, $\Theta = (20 \times 1.35)5$

$$= 27 \text{ \%}$$

+. Amount of water content $\qquad = 27 \times \dfrac{20}{100}$ cm

$$= 5.4 \text{ cm}$$

Example 27. Calculate the amount of water content in cm, in 0-60 cm soil profile from the following information:

Depth of soil layer (cm)	Moisture (%)	Bulk density (g/cc)
0 – 20	18.0	1.40
20 – 40	20.0	1.39
40 – 60	22.0	1.41

Solution :

Water content in 0 – 20 cm depth $= 18 \times 1.4 \times \dfrac{20}{100}$ cm

$\qquad\qquad\qquad = 5.04$ cm

Water content in 20 – 40 cm depth $= 20 \times 1.39 \times \dfrac{20}{100}$ cm

$\qquad\qquad\qquad = 5.56$ cm

Water content in 40 – 60 cm depth $= 22 \times 1.41 \times \dfrac{20}{100}$ cm

$\qquad\qquad\qquad = 6.20$ cm

+. Total water content $\qquad = (5.04 + 5.56 + 6.20)$ cm

$\qquad\qquad\qquad = 16.80$ cm

Example 28. The mass of soil sample taken from 0 – 30 cm soil layer is 200 g. the oven dry mass of the sample is 160 g. if the volume of the soil is 120 cc, calculate the moisture content in cm for 0 – 30 cm soil depth.

Solution :

Mass of wet soil, $\qquad\qquad M_t = 200$ g

Mass of oven dry soil, $\qquad M_s = 160$ g

Mass of water, $\qquad\qquad M_w = (200 - 160)$ g

$\qquad\qquad\qquad = 40$ g

Volume of water, $\qquad\qquad V_w = 40$ cc

Volume of soil, $\qquad\qquad V_t = 120$ cc

Volumetric moisture content $\qquad = \dfrac{V_w}{V_t} \times 100 \%$

Volumetric moisture content $\qquad = \dfrac{40}{120} \times 100 \%$

$$= 33.33 \%$$

Moisture content in 0 – 30 cm soil depth $= 33.33 \times \dfrac{30}{100}$ cm

$$= 10 \text{ cm}$$

Example 29. The maximum water holding capacity, field capacity and permanent wilting point for 0 – 30 cm soil depth are 46 %, 24 % and 11 %. If the bulk density of soil is 1.4 g/cc, calculate the (*a*) drainable water (*b*) available water and (*c*) unavailable water for 0 – 30 cm soil depth.

Solution :

(*a*) Drainable water

$$= \frac{(\text{Maximum water holding capacity - Field capacity}) \times \rho_b \times \text{depth of soil}}{100 \times \rho_w}$$

$$= \frac{(46 - 24) \times 1.4 \times 30}{100 \times 1} \text{ cm}$$

$$= 9.24 \text{ cm}$$

(*b*) Available water

$$= \frac{(\text{Field capacity - Permanent wilting point}) \times \rho_b \times \text{depth of soil}}{100 \times \rho_w}$$

$$= \frac{(24 - 11) \times 1.4 \times 30}{100 \times 1} \text{ cm}$$

$$= 5.46 \text{ cm}$$

(c) Unavailable water $= \dfrac{\text{Permanent wilting point} \times \rho_b \times \text{depth of soil}}{100 \times \rho_w}$

$$= \dfrac{11 \times 1.4 \times 30}{100 \times 1}\ \text{cm}$$

$$= 4.62\ \text{cm}$$

9.2.1. Net Irrigation Requirement (NIR)

Net quantity of water to be applied is equal to the moisture deficit in the soil *i.e.* the quantity required to fill the root zone to field capacity.

$$\text{NIR} = \sum_{i=1}^{n} \frac{\text{EC}_i - _i}{100} \times \text{ASG}_i \times \text{D}_i$$

Where,

FC_i = Field capacity of the *i*th layer on oven dry weight basis

$\cdot \S_i$ = Actual moisture content of the *i*th layer on oven dry weight basis at the time of irrigation

ASG_i= Apparent Specific Gravity (H" value of Bulk density in g/cc) of the *i*th layer

D_i = Depth of the *i*th layer

Example 30. Find out the quantity of irrigation water to be applied to 90 cm soil depth with following value of moisture:

Depth of soil water (cm)	Moisture content		Bulk density (g/cc)
	FC	¶§	
0 – 30	24.0	18.3	1.40
30 – 60	24.5	20.0	1.39
60 – 90	23.8	21.6	1.41

Solution :

Moisture deficit of Ist layer (0-30 cm) $= (24.0 - 18.3) \times 1.40 \times \dfrac{30}{100}$ cm

$$= 2.39 \text{ cm}$$

Moisture deficit of 2nd layer (30-60 cm) $= (24.5 - 20) \times 1.39 \times \dfrac{30}{100}$ cm

$$= 1.88 \text{ cm}$$

Moisture deficit of 3rd layer (60-90 cm) $= (23.8 - 21.6) \times 1.41 \times \dfrac{30}{100}$ cm

$$= 0.93 \text{ cm}$$

Total deficit $= (2.39 + 1.88 + 0.93)$ cm

$$= 5.2 \text{ cm}$$

+. Net depth of water to be applied is 5.2 cm.

Example 31. Find out the quantity of irrigation water to be applied to 90 cm soil depth with following value of moisture:

Depth of soil water (cm)	Moisture content		Bulk density (g/cc)
	FC	¶§	
0 – 30	10.08 cm	18.3 %	1.40
30 – 60	24.5 %	8.34 cm	1.39
60 – 90	10.07 cm	9.14 cm	1.41

Solution :

Moisture deficit of Ist layer (0-30 cm) $= (10.08 - 18.3 \times 1.40 \times \dfrac{30}{100})$ cm

$$= 2.39 \text{ cm}$$

Moisture deficit of 2nd layer (0-30 cm) $= (24.5 \times 1.39 \times \dfrac{30}{100} - 8.34)$ cm

$$= 1.88 \text{ cm}$$

Moisture deficit of 3rd layer (0-30 cm)	= (10.07 – 9.14) cm
	= 0.93 cm
Total deficit	= (2.39 + 1.88 + 0.93) cm
	= 5.2 cm

+. Net depth of water to be applied is 5.2 cm.

Example 32. If (*a*) 4 cm (*b*) 5 cm (*c*) 6 cm irrigation water is applied in the above exercise, how will water be distributed in 0-30, 30-60 and 60-90 cm soil profile ?

Solution :

(*a*) Deficit in the 0-30 cm layer is 2.39 cm

–   The layer will hold 2.39 water and rest (4 – 2.39) = 1.61 cm water will be drained away too the next layer.

–   30-60 cm layer has 1.88 cm deficit, the layer will hold the 1.61 cm water.

Thus, the distribution of 4 cm irrigation in 0-30 and 30-60 cm soil layer will be = 2.39 cm and 1.61 cm

(*b*) Deficit in the 0-30 cm layer is 2.39 cm

–   The layer will hold 2.39 water and rest (5 – 2.39) = 2.61 cm water will be drained away too the next layer.

–   30-60 cm layer has 1.88 cm deficit, the layer will hold the 1.88 cm water and rest (2.61 – 1.88) = 0.73 cm water will be drained away to the next layer.

Since, the distribution of 5 cm irrigation in 0-30, 30-60 and 60-90 cm soil layer will be = 2.39 cm, 1.88 cm and 0.73 cm

(*c*) Deficit in the 0-30 cm layer is 2.39 cm

–   The layer will hold 2.39 water and rest (6 – 2.39) = 3.61 cm water will be drained away too the next layer

–   30-60 cm layer has 1.88 cm deficit, the layer will hold the 1.88 cm water and rest (3.61 – 1.88) = 1.73 cm water will be drained away to the next layer.

- 60-90 cm layer has 0.93 cm deficit, the layer will hold the 0.93 cm water and rest $(1.73 - 0.93) = 0.80$ cm water will percolate down.

Thus, the distribution of 6 cm irrigation in 0-30, 30-60 and 60-90 cm soil layer will be = 2.39 cm, 1.88 cm and 0.93 cm. Rest 0.80 cm will percolate down beyond 90 cm depth.

Example 33. Find out the wetting depth, if 4.8 cm rainfall is received by soil with following values of moisture (assume that there is no run off loss):

Depth of soil water (cm)	Moisture content		Bulk density (g/cc)
	FC	¶§	
0 – 30	24.0	18.3	1.40
30 – 60	24.5	20.0	1.39
60 – 90	23.8	21.6	1.41

Solution :

Deficit in 0-30 cm $= (24.0 - 18.3) \times 1.40 \times \dfrac{30}{100}$ cm

$= 2.39$ cm

Deficit in 30-60 cm $= (24.5 - 20) \times 1.39 \times \dfrac{30}{100}$ cm

$= 1.88$ cm

After wetting, 60 cm soil depth, $(4.8 - 2.39 - 1.88)$ cm $= 0.53$ cm water percolated beyond 60 cm

Let depth 'd' be wetted by the water.

Hence, $0.73 = (23.8 - 21.6) \times 1.41 \times \dfrac{d}{100}$

or, $0.73 = 3.102 \times \dfrac{d}{100}$

or, $d = 0.73 \times \dfrac{100}{3.102}$

$= 23.53$ cm

Wetting depth of rain water $= (60 + 23.53)$ cm

$= 83.53$ cm

Example 34. The following data were obtained before irrigation. If 4 cm irrigation water is applied to 60 cm root zone depth, what is the effective depth of irrigation? How much water is wasted?

Depth of soil water (cm)	Moisture content		Bulk density (g/cc)
	FC	¶§	
0 – 20	24.0	20.4	1.40
20 – 40	24.5	22.1	1.39
40 – 60	23.8	23.0	1.41

Solution :

Requirement of irrigation at 0-20 cm depth $= (24 - 20.4) \times 1.4 \times \dfrac{20}{100}$ cm

$= 1.01$ cm

Requirement of irrigation at 20-40 cm depth $= (24.5 - 22.1) \times 1.39 \times \dfrac{20}{100}$ cm

$= 0.67$ cm

Requirement of irrigation at 40-60 cm depth $= (23.8 - 23.0) \times 1.41 \times \dfrac{20}{100}$ cm

$= 0.23$ cm

Hence requirement of irrigation water at 0-60 cm depth

$= (1.01 + 0.67 + 0.23)$ cm

$= 1.91$ cm

Thus, effective depth of irrigation water is 1.91 cm and rest 2.09 cm is percolated beyond 60 cm.

9.2.2. Gross Irrigation Requirement (GIR)

The total amount of water applied through irrigation is termed as gross irrigation requirement. In other words, it is net irrigation requirement plus losses in water application and other losses. The gross irrigation requirement can be determined for a field, for a farm, fpr a command area or for an irrigation project, depending on the need, by considering the appropriate losses at various stages of water conveyance and distribution.

$$\text{Gross irrigation requirement (GIR)} = \frac{\text{Net irrigation requirement}}{\text{Field efficiency of the system}}$$

Example 35. If the net irrigation requirement is 4.10 cm and field efficiency is 70%, calculate the gross irrigation requirement.

Solution :

$$\text{Gross irrigation requirement} = 4.10 \times \frac{100}{70} \text{ cm}$$

$$= 5.86 \text{ cm}$$

9.2.3. Scheduling of Irrigation

Irrigation scheduling indicates when and how much irrigation water is to be applied. Several approaches for scheduling irrigation have been used by different people.

9.2.3.1. *Soil moisture depletion approach*

In this approach, when soil moisture in a specified root zone is depleted to a particular level, it is replenished by water.

Example 36. Irrigation to upland rice is scheduled at 25% depletion of available soil moisture. Field capacity of soil in the effective root zone depth of 60 cm is 24% with a permanent wilting point of 10%. At which soil moisture content irrigation is to be scheduled? Also calculate the depth of water, if bulk density of the soil is 1.4 g/cc.

Solution :

Field capacity (FC)   = 24 %

Permanent wilting point (PWP) = 10 %

∴ Available moisture   = (24 – 10) %

                                           = 14 %

Given that, irrigation is to be done at 25 % depletion of available water

∴ 25% of 14%   = $\left[14 \times \dfrac{25}{100}\right]$ %

                                           = 3.5 %

∴ Irrigation should be applied at = (24 – 3.5)%

                                           = 20.5 % soil moisture

and Net depth of irrigation water = $3.5 \times 1.4 \times \dfrac{60}{100}$ cm

                                           = 2.94 cm

Alternative way

Available water   = $(2.4 – 10) \times 1.4 \times \dfrac{60}{100}$ cm

                                           = 11.76 cm

*. Irrigation is to be applied at 25% depletion of available water, net depth

   of irrigation water   = $11.76 \times \dfrac{25}{100}$ cm

                                           = 2.94 cm

   Water content at 0-60 cm at FC = $24 \times 1.4 \times \dfrac{60}{100}$ cm

                                           = 20.16 cm

∴ Irrigation should be applied at water content

                                           = (20.16 – 2.94) cm

                                           = 17.22 cm per 60 cm

+. Water content per 100 cm = $17.22 \times \dfrac{60}{100}$ cm = 28.7 cm

i.e. volumetric moisture content = 28.7 %

+. Gravimetric moisture content at 25 % depletion of available water

$$= \frac{28.7}{1.4}\%$$

$$= 20.5\%$$

9.2.3.2. *IW/CPE approach*

In this approach, a known amount of irrigation water (IW) is applied when cumulative pan evaporation (CPE) reaches a predetermined level. Scheduling irrigation at an IW/CPE ratio 1.0 with 4 cm irrigation means that 4 cm irrigation water is applied when CPE reaches 4 cm.

Example 37. Calculate the cumulative pan evaporation required for scheduling irrigation at IW/CPE ratio (*a*) 0.8 (*b*) 1.0 and (*c*) 1.2 with 4.0 cm irrigation water.

Solution :

(*a*) $\qquad \dfrac{IW}{CPE} = 0.8$

or $\qquad \dfrac{5.0\,cm}{CPE} = 0.8$

or $\qquad CPE = \dfrac{5}{CPE}\ cm$

or $\qquad CPE = 6.25$ cm

(*b*) $\qquad \dfrac{IW}{CPE} = 1.0$

or $\qquad \dfrac{5.0\ cm}{CPE} = 1.0$

or $\qquad$ CPE $\quad = \dfrac{5}{10}$ cm

or $\qquad$ CPE $\quad = 5.0$ cm

(c) $\qquad \dfrac{IW}{CPE} = 1.2$

or $\qquad \dfrac{5.0 \text{ cm}}{CPE} = 1.2$

or $\qquad$ CPE $\quad = \dfrac{5}{1.2}$ cm

or $\qquad$ CPE $\quad = 4.17$ cm

Example 38. In a wheat crop, 6 irrigation was scheduled at IW/CPE ratio 0.9. If 3 cm effective rainfall comes before desired CPE level, calculate the CPE and depth of irrigation.

Solution :

$$\dfrac{IW}{CPE} = 0.9$$

or $\qquad \dfrac{6.0 \text{ cm}}{CPE} = 0.9$

or $\qquad$ CPE $\quad = \dfrac{6.0}{0.9}$ cm

or $\qquad$ CPE $\quad = 6.67$ cm

How should we adjust the effective rainfall?

Here, different people have different opinion:

(a) Group I: Don't consider rainfall. Apply the pre-fixed depth of irrigation.

(b) Group II: Apply water deducting the effective rainfall.

(c) Group III: Convert the effective rainfall to equivalent CPE and wait till the desired CPE comes.

173

Now the question is –

(a) If we receive 6 cm rainfall just before irrigation or a day before irrigation, should we again apply 6 cm irrigation? Thus, opinion of this group is not acceptable.

(b) Deducting the effective rainfall from the pre-fixed depth of irrigation is very logical. However, some problem may arise again. In this example, required depth of irrigation will be 6 – 3 = 3 cm. In a small experimental plot, it may be possible to apply 3 cm irrigation water. Is it possible to apply 3 cm irrigation through flooding in a large field? Thus, a reasonable depth (say 4 cm) should be considered even if there is 1 cm loss of irrigation water.

(c) It is mathematically correct. Thus, in this example, irrigation should be applied when 6.67 + 3.33 = 10.0 cm CPE is reached. But problem arises again. Water is freely available to plants to a certain point (near FC) in the available range and subsequently availability is decreased with decrease in soil moisture. Thus, if irrigation is delayed to adjust this 3.0 cm effective rainfall (that 3 cm rainfall may not raise the water level to FC), the crop will be exposed to more water stress periods.

Thus, opinion of the second group looks logical.

9.2.4. Volume of Irrigation Water

Although irrigation water is recommendation in terms of depth, in field, we have to convert it to volume to apply water in a particular plot or area. Thus, finally irrigation water has to be applied in terms of volume. The volume of irrigation water is calculated as follows:

$$q = a \times d$$

where,

q = quantity of water needed (m^3)

a = area to be irrigated (m^2)

d = depth of water (m)

Example 39. Calculate the volume of water required to apply 6 cm water in a 100 m² plot.

Solution :

Area of the plot	$= 100 \text{ m}^2$
Depth of irrigation water	$= 6 \text{ cm}$
	$= 0.6 \text{ m}$
∴ Volume of irrigation water	$= 100 \text{ m}^2 \times 0.6 \text{ m}$
	$= 60 \text{ m}^3$
	$= 60 \times 1,000 \text{ lit [since, } 1 \text{ m}^3 = 1,000 \text{ lit]}$

Volume of irrigation water (lit) = Area (m²) × Depth of irrigation (m) × 1000

$$= 60,000 \text{ lit}$$

Example 40. A 2,000 m² wheat plot I to be irrigated with 6 cm water through STW. If the discharge of the pump is 8000 lit/hour, calculate the pumping time, if field efficiency is 80%.

Solution :

Net irrigation requirement	$= 6 \text{cm}$
Field efficiency	$= 80\%$
∴ Gross irrigation requirement	$= 6 \times \dfrac{100}{80} \text{ cm}$
	$= 7.5 \text{ cm}$
∴ Volume of irrigation water	$= 2000 \times \dfrac{7.5}{100} \times 1000 \text{ lit}$
	$= 1,50,000 \text{ lit}$

Again, 8000 lit is delivered by the pump in

$$= 1 \text{ hour}$$

∴ 1,50,000 lit is delivered by the pump in $= \dfrac{1,50,000}{8,000} \text{ hour}$

$$= 18.75 \text{ hour}$$

$$= 18 \text{ hour } 45 \text{ minutes}$$

Required duration of irrigation (hour) = $\dfrac{\text{Volume of water required (lit)}}{\text{Discharge (lit/hour)}}$

Example 41. Calculate the required to irrigate 1 ha land from the following information, if irrigation water is to be applied at 40 % depletion of available water and field efficiency of the system is 80 %

(a) Field capacity = 23.5%

(b) Permanent wilting = 11%

(c) Bulk density = 1.4 g/cc

(d) Depth of irrigation = 90 cm

(e) Discharge = 100 lit/sec

Solution :

Available water content in 90 cm depth = $(23.5 - 11) \times 1.4 \times \dfrac{90}{100}$ cm

$$= 15.75 \text{ cm}$$

Since irrigation is to be done at 40% depletion of available water,

+. Net depth if irrigation water $\qquad = 15.75 \times \dfrac{40}{100}$ cm

$$= 6.30 \text{ cm}$$

Given that, field capacity $\qquad = 80\%$

+.   Gross depth of irrigation $\qquad = 6.30 \times \dfrac{100}{80}$ cm

$$= 7.875 \text{ cm}$$

+.   Volume of water required to irrigated 1ha area

$$= 10,000 \times \dfrac{7.875}{100} \times 1,000 \text{ lit}$$

$$= 7,87,500 \text{ lit}$$

Discharge of the pump $\qquad = 100 \text{ lit/sec}$

+. Time required to irrigate the area $= \dfrac{7,87,500}{100} \text{ sec}$

$$= 7875 \text{ sec}$$

$$= 2 \text{ hour } 11 \text{ minutes } 15 \text{ sec}$$

Example 42. A pump discharge at the rate of 60,000 lit/hour and works for eight hours each day. Estimate the area commanded by the water discharged, if the average depth of irrigation is 6 cm and irrigation interval is 15 days.

Solution:

Total discharge in 15 days $= \dfrac{60,000 \times 8 \times 15}{1,000} \text{ m}^3$

$$= 7,200 \text{ m}^3$$

Water required to cover 1 ha to a depth of 6 cm

$$= 10,000 \times \dfrac{6}{100}$$

$$= 600 \text{ m}^3$$

Area command by the pump $= \dfrac{\text{Water available}}{\text{Water required per unit area}}$

$$= \dfrac{7,200}{600}$$

$$= 12 \text{ ha}$$

Example 43. The moisture content of a soil (having bulk density 1.4 g/cc at 80 cm effective root zone) at FC and PWP are 24% and 11%, respectively. The field is to be irrigated at 25% depletion of available water. If rate of evaporation is 0.5 cm/day and irrigation efficiency is 70%, calculate the net and gross depth of irrigation and irrigation interval.

Solution:

Available soil moisture in the root zone $= (24 - 11) \times 1.4 \times \dfrac{80}{100}$ cm

$= 14.56$ cm

Net depth of irrigation to be applied $= 14.56 \times \dfrac{25}{100}$ cm

$= 3.64$ cm

Gross depth of irrigation to be applied $= 3.64 \times \dfrac{100}{70}$ cm

$= 5.20$ cm

Irrigation interval $= \dfrac{5.2}{0.5}$

$= 10.4$

≈ 10 days

9.2.5. Irrigation Efficiencies

Entire quantity of applied irrigation water is never stored in the root zone even under best management practices. Irrigation efficiency shows how efficiently irrigation water is conveyed, applied, stored and distributed in the field.

9.2.5.1. Water Conveyance Efficiency (E_c)

This term is used to measure the efficiency of water conveyance systems associated with the canal networks, water courses and field channels from the well to the individual fields. It is expressed as follows:

$$E_c = \frac{W_f}{W_d} \times 100$$

Where,

E_c = Water conveyance efficiency (%)

W_f = Water delivered to the irrigated plot (at the field supply channel)

W_d = Water delivered from the source

Example 44. 5,00,000 lit water was delivered from the pump. However, only 3,60,000 lit water reached the field supply channel. Calculate the water conveyance efficiency.

Solution:

$$\text{Water conveyance efficiency} = \frac{3,60,000}{5,00,000} \times 100\%$$

$$= 72 \%$$

Example 45. A stream of 125 lit/sec was diverted from a canal and 90 lit/sec was delivered to the field supply channel. Calculate the water conveyance efficiency.

Solution:

$$\text{Water conveyance efficiency} = \frac{95}{125} \times 100\% = 72\%$$

9.2.5.2. *Water Application Efficiency* (E_a)

It is expressed as follows:

$$E_a = \frac{W_s}{W_f} \times 100$$

Where,

E_a = Water application efficiency (%)

W_s = Water stored in the root zone of the plants, and

W_f = Water delivered to the field (at the supply channel)

Example 46. 3,60,000 lit water was delivered to the field. However, only 2,70,000 lit water reached the root zone. Calculate the water application efficiency.

Solution:

$$\text{Water application efficiency} = \frac{2,70,000}{3,60,000} \times 100\%$$

$$= 75\%$$

Example 47. 8 cm water was delivered to a field. However, only 6 cm water reaches the root zone. Calculate the water application efficiency.

Solution:

$$\text{Water application efficiency} = \frac{6}{8} \times 100\%$$

$$= 75\%$$

9.2.5.3. *Project Efficiency* (E$_p$)

Project Efficiency (E$_p$) is the ratio of average depth of water stored during irrigation (W$_s$) to water delivered from the source (W$_d$).

$$E_p = \frac{W_s}{W_d} \times 100$$

Where,

E$_p$ = Project efficiency (%)

W$_s$ = Water stored in the root zone of the plants, and

W$_d$ = Water delivered from the source

Thus,
$$E_p = \frac{E_c \times E_a}{100}$$

Example 48. 5,00,000 lit water was delivered from a pump. However, only 3,60,000 lit water reached the field supply channel and 2,70,000 lit reached the root zone. Calculate the project efficiency.

Solution:

Project efficiency $= \dfrac{2,70,000}{5,00,000} \times 100\%$

$= 54\%$

Alternatively

$$E_c = \dfrac{3,60,000}{5,00,000} \times 100\%$$

$$= 72\%$$

$$E_a = \dfrac{2,70,000}{3,60,000} \times 100\%$$

$$= 75\%$$

+.

$$E_p = \dfrac{E_c \times E_a}{100}$$

$$= \dfrac{72 \times 75}{100}\%$$

$$= 54\%$$

9.2.5.4. *Water Storage Efficiency*

The water storage efficiency refers to the ratio of water stored in root zone during irrigation to the water needed prior to irrigation. It is expressed as:

$$E_s = \dfrac{W_s}{W_n} \times 100$$

Where, E_s = Water storage efficiency (%)

W_s = Water stored in the root zone after irrigation, and

W_n = Water needed in root zone prior to irrigation

Example 49. 4,50,000 lit water was needed in a crop field to reach field capacity level. However, only 2,70,000 lit water was stored in the root zone. Calculate the water storage efficiency.

Solution:

$$\text{Water storage efficiency} = \frac{2,70,000}{4,50,000} \times 100\%$$

$$= 60\%$$

Example 50. 10 cm water was needed in a crop field to reach field capacity level in the root zone. However, only 6 cm water was stored in the root zone. Calculate the water storage efficiency.

Solution:

$$\text{Water storage efficiency} = \frac{6}{10} \times 100\%$$

$$= 60\%$$

Example 51. Soil moisture contents in different depths just before irrigation and 2 days after irrigation in a plot are given in the following Table. The depth of water applied to plot as measured at the plot head was 20 cm. Work out the irrigation water application efficiency. (Assume that PET for two days following irrigation to be 1 cm).

Soil depth (cm)	Soil moisture before irrigation (%)	Soil moisture 2 days after irrigation (%)	Bulk density (g/cc)
0-30	12.0	25.0	1.48
30-60	18.2	25.2	1.50
60-90	18.2	26.2	1.50
90-120	20.2	25.2	1.48
120-150	20.2	25.6	1.52

Solution :

Soil depth (cm)	Moisture retained in the profile (cm)
0-30	[(25.0 – 12.0)/100] ×1.48 ×30 = 5.77
30-60	[(25.2 – 18.2)/100] ×1.50 ×30 = 3.15
60-90	[(26.2 – 18.2)/100] ×1.50 ×30 = 3.60
90-120	[(25.2 – 20.2)/100] ×1.48 ×30 = 2.22
120-150	[(25.6 – 20.2)/100] ×1.52 ×30 = 2.46
Total	17.20

Thus, moisture retained in the profile $\quad$ = 17.20 cm
Water loss during 2 days $\quad$ = 1.00 cm
+. water stored in root zone after irrigation (W_s) $\quad$ = (17.20 + 1.00) cm
$\quad$ = 18.20 cm

Water delivered to the plot (W_f) $\quad$ = 20 cm

Water application efficiency, E_a $\qquad = \dfrac{W_s}{W_f} \times 100\%$

$$= \frac{18.20}{20.00} \times 100\%$$

$$= 91\%$$

Example 52. Soil moisture contents in different depths just before irrigation and 2 days after irrigation in a plot are given in the following Table. The depth of water applied to plot was calculated to reach field capacity level. Calculate the water storage efficiency. Assume that ET loss is negligible during these two days.

Soil depth (cm)	Soil moisture before irrigation (%)	Field capacity (%)	Soil moisture 2 days after irrigation (%)	Bulk density (g/cc)
0-30	12.0	25.0	25.0	1.48
30-60	18.2	25.2	25.2	1.50
60-90	18.2	26.2	26.2	1.50
90-120	20.2	25.2	22.6	1.48
120-150	20.2	25.6	20.2	1.52

Solution :

Soil depth (cm)	Water required (cm)	Water retained in the profile (cm)
0-30	$\dfrac{(25.0 - 12.0) \times 1.48 \times 30}{100} = 5.77$	$\dfrac{(25.0 - 12.0) \times 1.48 \times 30}{100} = 5.77$
30-60	$\dfrac{(25.2 - 18.2) \times 1.50 \times 30}{100} = 3.15$	$\dfrac{(25.2 - 18.2) \times 1.50 \times 30}{100} = 3.15$
60-90	$\dfrac{(26.2 - 18.2) \times 1.50 \times 30}{100} = 3.60$	$\dfrac{(26.2 - 18.2) \times 1.50 \times 30}{100} = 3.60$
90-120	$\dfrac{(25.2 - 20.2) \times 1.48 \times 30}{100} = 2.22$	$\dfrac{(22.6 - 20.2) \times 1.48 \times 30}{100} = 1.07$
120-150	$\dfrac{(25.6 - 20.2) \times 1.52 \times 30}{100} = 2.46$	$\dfrac{(20.2 - 20.2) \times 1.52 \times 30}{100} = 0$
Total	17.20	13.59

$+.$ Water storage efficiency $= \dfrac{13.59}{17.20} \times 100\%$

$$= 79.01\%$$

9.2.5.5. *Water Distribution Efficiency*

Water distribution efficiency indicated the extent to which water is uniformly distributed along the run. It is defined as:

$$E_d = \left[1 - \frac{\hat{y}}{d}\right] \times 100$$

Where,

E_d = Water distribution efficiency (%)

d = Average depth of water stored along the run during the irrigation

$\hat{y}$ = Average numerical deviation from d

Example 53. If the penetration depth of irrigation water after irrigation from head end to the tail end are 150 cm, 145 cm, 140 cm, 130 cm, 120 cm, 105 cm and 85 cm, calculate the water distribution efficiency.

Solution:

$$\text{Average storage depth, } d = \frac{150 + 145 + 140 + 130 + 120 + 105 + 85}{7} \text{ cm}$$

$$= 125 \text{ cm}$$

Deviation from mean*:

$$(150 - 125) = 25 \text{ cm}$$
$$(145 - 125) = 20 \text{ cm}$$
$$(140 - 125) = 15 \text{ cm}$$
$$(130 - 125) = 5 \text{ cm}$$
$$(125 - 120) = 5 \text{ cm}$$
$$(125 - 105) = 20 \text{ cm}$$
$$(125 - 85) = 40 \text{ cm}$$

[*always consider positive values]

+. Mean deviation, $w = \dfrac{25 + 20 + 15 + 5 + 5 + 20 + 40}{7} \text{ cm}$

$$= \frac{130}{7} \text{ cm}$$

$$= 18.57 \text{ cm}$$

+. Water distribution efficiency, $E_d = \left[1 - \dfrac{\hat{y}}{d}\right] \times 100\%$

$$= \left[1 - \frac{18.57}{125}\right] \times 100\%$$

$$= 85.14\%$$

9.2.5.6. *Water Use Efficiency*

(*i*) ***Crop Water Use Efficiency (CWUE)*** **:** It is the ratio of crop yield (Y) to the amount of water depleted by the crop in the process of evaporation (ET). Generally, water use efficiency indicates Crop Water Use Efficiency.

$$CWUE = \frac{Y}{ET}$$

Example 53. Calculate the CWUE of mustard, if productivity is 1200 kg/ha and ET is 15.5 cm.

Solution :

$$CWUE \text{ of mustard} = \frac{1200}{15.5} \text{ kg/ha-cm}$$

$$= 77.42 \text{ kg/ha-cm}$$

Example 54. If the grain yield of wheat is 4.5 t/ha and ET is 400 mm, calculate the water use efficiency.

Solution :

Grain yield $= 4.5 \times 1000$ kg

$= 4500$ kg

ET $= 400$ mm

$= 40$ cm

Water use efficiency $= \dfrac{4,500}{40} \text{ kg/ha-cm}$

$= 112.5 \text{ kg/ha-cm}$

(*ii*) ***Field Water Use Efficiency (FWUE)*** **:** It is the ratio of crop yield (Y) to the total amount of water used in the field (WR).

$$\text{Field water use efficiency (FWUE)} = \frac{Y}{WR}$$

Example 53. Calculate the FWUE of mustard, if productivity is 1200 kg/ha and WR is 28.2 cm.

Solution :

FWUE of mustard $= \dfrac{1,200}{28.2}$ kg/ha-cm

$= 42.55$ kg/ha-cm

9.2.6. Design Irrigation Frequency

In designing irrigation systems, the irrigation frequency to be used is the time (in days) between twi irrigations in the period of highest consumptive use of the crop growth. Irrigation frequency depends on how fast soil moisture is extracted when a crop is transpiring at its maximum rate. The average moisture-use rate during this period is used to plan irrigation systems. For an irrigation system to be adequate, it must have sufficient capacity to supply the water required during this period. The design irrigation frequency may be computed as follows:

$$\text{Design frequency (days)} = \frac{M_{fc} - M_{bi}}{P_{cu}}$$

Where,

M_{fc} = Field capacity of the soil in the effective crop root zone

M_{bi} = Moisture content of the same zone at the time of starting of irrigation

P_{cu} = Peak period moisture use rate of crop

Example 56. A 26 ha rice land should be irrigated by a pump. The recommended irrigation management practice is to apply 5 cm irrigation at 3 days after disappearance of ponded water (DADPW). The peak period ET is 4.5 mm/day and percolation loss is 0.4 mm/day. If the water application efficiency is 75% and the pump is run 8 hours a day, calculate the minimum irrigation period and required capacity of the irrigation system.

Solution:

Peak period ET $= 4.5$ mm/day

Percolation loss $= 0.5$ mm/day

The maximum portable daily loss of water

$$= (4.5 + 0.5) \text{ mm} = 5.0 \text{ mm}$$

Hence, time to lose 5 cm water from the field

$$= \frac{50}{5} \text{ days}$$

$$= 10 \text{ days}$$

Since, irrigation is to be done at 3 DADPW, next irrigation should be done after

$$= (10 + 3) \text{ days}$$

$$= 13 \text{ days}$$

This means that, the minimum days required to re-irrigate the field is 13 days

Thus minimum irrigation period $= 13$ days

+. The pump must have the capacity to irrigate $= \dfrac{26}{13}$

$$= 2 \text{ ha per day in 8 hours.}$$

*. The irrigation efficiency is 75%,

+. Pumping depth/application $= 5 \times \dfrac{100}{75}$

$$= 6.67 \text{ cm}$$

Thus, required capacity of the irrigation system

$$= \frac{(6.67/100) \text{ m} \times 20{,}000 \text{ m}^2}{80 \times 60 \times 60 \text{ sec}}$$

$$= \frac{1{,}334 \text{ m}^3}{28{,}800 \text{ sec}}$$

$$= \frac{1{,}334 \times 1{,}000 \text{ lit}}{28{,}800 \text{ sec}}$$

$$= 46.32 \text{ lit/sec}$$

9.2.7. Effective Rainfall (ER)

Effective rainfall is the amount of rainfall reaching the field during the growing period of crop that is available to meet the consumptive use of crop. It is the rainfall excluding deep percolation beyond the root zone and surface run off..

Effective rainfall = Rainfall – (deep percolation + surface runoff)

Example 57. The moisture content of root zone of two nearby plot just before rainfall are 17 cm (irrigated plot) and 12 cm (rainfed plot) and the field capacity of root zone is 20 cm. Bunds are provided around the plots to check run off. Calculate the effective rainfall and deep percolation in both the plots, if (*a*) 2 cm (*b*) 3 cm (*c*) 5 cm (*d*) 10 cm rainfall is received.

Solution:

Water deficit of irrigated plot   = (20 – 17) cm = 3 cm

Water deficit of rainfed plot   = (20 – 12) cm = 8 cm

(*a*) If 2 cm rainfall is receive, ER for both the plots will be = 2 cm and no deep percolation

(*b*) If 3 cm rainfall is receive, ER for both the plots will be = 3 cm and no deep percolation

(*c*) If 5 cm rainfall is receive, ER for

Irrigated plot	= 3 cm
Rainfed plot	= 5 cm
Deep percolation in irrigated plot	= (5 – 3) cm = 2cm
Deep percolation in rainfed plot	= Nil

(d) If 10 cm rainfall is receive, ER for

Irrigated plot	= 3 cm
Rainfed plot	= 8 cm
Deep percolation in irrigated plot	= (10 – 3) cm = 7cm
Deep percolation in rainfed plot	= (10 – 3) cm = 2cm

Example 58. The moisture content of two nearby plots just before rainfall is given below:

	Irrigated plot			Rainfed plot		
Soil depth	Moisture content before rainfall (%)	FC (%)	BD (g/cc)	Moisture content before rainfall (%)	FC (%)	BD (g/cc)
0-30 cm	20.7	24.0	1.36	14.1	24.0	1.36
30-60 cm	22.2	24.1	1.37	16.2	24.1	1.37
60-90 cm	23.3	24.2	1.37	19.3	24.2	1.37

Bunds are provided around the plots to check run off. Root zone depth is 90 cm. Calculate the effective rainfall and deep percolation in both the plots, if 8 cm rainfall is received.

Solution:

Soil depth	Irrigated plot	Rainfed plot
	Moisture deficit (cm)	Moisture content before rainfall (%)
0-30 cm	(24.0 – 20.7) ×1.36 ×30/100 = 1.35	(24.0 – 14.1) ×1.36 ×30/100 = 4.04
30-60 cm	(24.1 – 22.2) ×1.37 ×30/100 = 0.78	(24.1 – 16.2) ×1.37 ×30/100 = 3.25
60-90 cm	(24.2 – 23.3) ×1.37 ×30/100 = 0.37	(24.2 – 19.3) ×1.37 ×30/100 = 2.01
Total	2.50 cm	9.30 cm

In case if irrigated plot, ER	= 2.50 cm	
& Deep percolation	= (8 – 2.5) cm	
	= 5.5 cm	
In case of rainfed plot, ER	= 8.00 cm	
And deep percolation	= Nil	

Example 59. Calculate the ER between 1-15 May, 2013 from the following observation if the storage capacity of root zone is 10.0 cm and initial moisture at the root zone is 8.5 cm.

Date	Rainfall (cm)	ET (cm)
30.04.2013	-	-
01.05.2013	-	0.41
02.05.2013	-	0.42
03.05.2013	2.0	0.29
04.05.2013	-	0.33
05.05.2013	-	0.39
06.05.2013	3.0	0.30
07.05.2013	-	0.35
08.05.2013	-	0.38
09.05.2013	12.0	0.2
10.05.2013	-	0.25
11.05.2013	-	0.39
12.05.2013	-	0.42
13.05.2013	-	0.45
14.05.2013	3.0	0.31
15.05.2013	-	0.36

Solution :

Date	Rainfall (cm)	ET (cm)	Balance (cm)	Surplus (cm)	ER (cm)
30.04.2013	-	-	8.50	-	-
01.05.2013	-	0.41	8.09		
02.05.2013	-	0.42	7.67		
03.05.2013	2.0	0.29	9.38	-	2.0
04.05.2013	-	0.33	9.05		
05.05.2013	-	0.39	8.66		
06.05.2013	3.0	0.30	10.0	1.36	1.64

[Table Contd.

Contd. Table]

Date	Rainfall (cm)	ET (cm)	Balance (cm)	Surplus (cm)	ER (cm)
07.05.2013	-	0.35	9.65		
08.05.2013	-	0.38	9.27		
09.05.2013	12.0	0.2	10.00	11.07	0.93
10.05.2013	-	0.25	9.75		
11.05.2013	-	0.39	9.36		
12.05.2013	-	0.42	8.94		
13.05.2013	-	0.45	8.49		
14.05.2013	3.0	0.31	10.00	1.18	1.82
15.05.2013	-	0.36	9.64		
TOTAL	**20.0 cm**			**13.61 cm**	**6.39 cm**

Thus ER between 1 – 15 May, 2013 = 6.36 cm

9.2.8. Irrigation systems terminology

Gross Command Area (GCA): It is the total area, bounded with irrigation boundary of a project, which can be economically irrigated without considering the limitation of the quantity of available water. It includes the cultivable as well as the non-cultivable area (roads, buildings, non- cultivable Area etc.).

Cultivable Command Area (CCA): It is the part of GCA in which cultivation is possible. It is the actual area under irrigation with crops.

Intensity of Irrigation (II): It is the actual area irrigated from an outlet. It is the ratio of actually irrigated area during crop season to the net CCA or the percentage CCA proposed to be irrigated seasonally.

Time Factor: The time factor is the number of dats the canal has actually run to number of days the canal was supposed to run. A time factor 15/20 means the canal actually runs 15 days although it is supposed to run for 20 days.

Base Period (B): It refers to the entire duration of crop in days from irrigation for preparatory cultivation to the last irrigation.

Delta (Δ): Delta refers to the total depth of water required by a crop during its duration in the field. If the crop of 100 days is irrigated at 10 days interval with 8 cm depth of water at each irrigation, the delta for crop is 80 cm.

Duty of water (D): Duty represents the irrigating capacity of a unit of water. It is the relation between the area of a crop irrigated and quantity of irrigation water required during the entire period of growth of that crop.

The duty of water is reckoned in the following four ways –

(i)   by number of ha that 1 cumec (*i.e.* 1 m^3/sec) of irrigation can irrigate during base period

e.g. 1500 ha/cumec

(ii) by total depth of water (or delta)

e.g. 1.5 m

(iii) by number of hectares that can be irrigated by million cubic metre of stored water. This system is used in case of tank irrigation.

e.g. ha/Mm^3

(iv) by number of ha-metre expended per hectare irrigation. This system is also used in case of tank irrigation.

e.g. ha-m/ha

Relation between duty and delta

If 1 cumec (*i.e.* m^3/sec) water is applied for base period B, the volume of water (V) applied to the crop during B days = $1 \times 60 \times 60 \times 24 \times B = 86400$ B m^3

Let, duty of water $= D$ ha/cumec

+. $$\Delta = \frac{86,400 \text{ B m}^3}{D \text{ ha}}$$

$$= \frac{86,400 \text{ B m}^3}{10,000 \times \text{D m}^2}$$

$$= \frac{8.64 \text{ B}}{\text{D}} \text{ m}$$

$$= \frac{864 \text{ B}}{\text{D}} \text{ m}$$

[1 cumec-day = 8.64mha-m = 86400 m³]

Example 60. Total depth of water application for a crop of 130 days base period is 110 cm. What is the duty of water?

Solution:

Here,   B   = 130 days

           Δ = 110 cm

           D   = ?

*. Δ $= \dfrac{864 \text{ B}}{\text{D}}$ cm

or $110 = \dfrac{864 \times 130}{\text{D}}$

or $\text{D} = 864 \times \dfrac{130}{100}$

           = 1021.1 ha/cumec

Example 61. Calculate the delta of a crop with a base period of 120 days and 1200 ha/cumec duty.

Solution:

Here,   B   = 120 days

           Δ = ?

           D   = 1200 ha/cumec

*.
$$\Delta = \frac{864\ B}{D}\ \text{cm}$$

or
$$\Delta = 864 \times \frac{120}{1,200}\ \text{cm}$$

$$= 86.4\ \text{cm}$$

Example 62. Gross command area of an irrigation canal is 40,000 ha. Cultivable irrigated area is 60%. Intensity of irrigation is 20 and 80% for *kharif* and *rabi*, respectively. What is the discharge required at the head of the canal, if the duty at its head is 400 and 1,100 ha/cumec for *rabi* and *kharif*, respectively?

Solution:

Cultivable irrigated area
$$= 40,000 \times \frac{60}{100}$$
$$= 24,000\ \text{ha}$$

Kharif area
$$= 24,000 \times \frac{20}{100}$$
$$= 4,800\ \text{ha}$$

Rabi area
$$= 24,000 \times \frac{80}{100}$$
$$= 19,200\ \text{ha}$$

Discharge requirement at head in *kharif*
$$= \text{Area under irrigation/duty}$$
$$= \frac{4,800}{400}$$
$$= 12.0\ \text{cumec}$$

Discharge requirement at head in *rabi*
$$= \text{Area under irrigation/duty}$$
$$= \frac{19,200}{1,100}$$
$$= 7.45\ \text{cumec}$$

Example 63. Cultivable command area of a canal is 2,000 ha. Intensity of irrigation for sugarcane and *rabi* vegetables are 25% and 35% with duty 1,500 and 1,200 ha/cumac, respectively. Calculate the required discharge at head end, if capacity factor is 0.80.

Solution:

$$\text{Irrigated area under sugarcane} = 1{,}500 \times \frac{25}{100}$$

$$= 375 \text{ ha}$$

$$\text{Irrigated area under sugarcane} = 1{,}200 \times \frac{35}{100}$$

$$= 420 \text{ ha}$$

Discharge requirement at head in sugarcane = Area under irrigation/duty

$$= \frac{375}{1{,}500}$$

$$= 0.25 \text{ cumec}$$

Discharge requirement at head in *rabi* vegetables= Area under irrigation/duty

$$= \frac{420}{1{,}200}$$

$$= 0.35 \text{ cumec}$$

*. Sugarcane requires irrigation for the whole year, during *rabi*, both sugarcane and vegetables will be requiring water

+. To meet the demand of both the crops, the discharge = (0.25 m + 0.35) cumec

$$= 0.60 \text{ cumec}$$

Again, it is given that, capacity factor = 0.80

+. Design discharge

$$= \frac{0.60}{0.80} \text{ cusec}$$

$$= 0.75 \text{ cusec}$$

Example 64. If peak demand is 20% more than the average requirement in the above example, calculate the design discharge.

Solution:

*. Peak demand is 20% more than the average requirement

+. Design discharge will be $= 0.75 \times \dfrac{0.60}{0.80}$ cumec

$$= 0.90 \text{ cumec}$$

Example 65. Water is released at the rate of 10 cumec at the head end. Calculate the area that can be irrigated, if duty at field is 100 ha/cumec and conveyance loss is 15%

Solution:

Discharge at head	= 10 cumec
Duty	= 100 ha/cumec
+. Area that can be irrigated	= Discharge × Duty
	= 10 × 100
	= 1,000 ha

*. 15% water is lost as conveyance loss,

+. Actual area that can be irrigated $= 1,000 \times 0.85$

$$= 850 \text{ ha}$$

Example 66. In a 500 ha canal command area, kharif rice – *toria* cropping sequence is followed. The intensity of *kharif* rice and *toria* are 80% and 50%, respectively. The duty of these crops at head end of water course are 700 and 1,300 ha/cumec. Find the discharge required, if peak demand is 20% more than the average requirement.

Solution:

Irrigated area under *kharif* rice $= 500 \times \dfrac{80}{100}$

$$= 400 \text{ ha}$$

Irrigated area under *toria* $= 500 \times \dfrac{50}{100}$

$= 250$ ha

Rabi area $= 24,000 \times \dfrac{80}{100}$

$= 19,200$ ha

Discharge requirement at head in *rabi*

$$= \dfrac{\text{Area under irrigation}}{\text{Duty}}$$

$$= \dfrac{400}{700}$$

$= 0.57$ cumec

Discharge requirement at head for *toria*

$$= \dfrac{\text{Area under irrigation}}{\text{Duty}}$$

$$= \dfrac{250}{1,300}$$

$= 0.19$ cumec

Since, *kharif* rice and *toria* are grown in different season, they will not compete for water. Hence, the discharge should be designed to meet the highest discharge, *i.e.*, in this case, to meet the demand of *kharif* rice.

+. Design discharge $= 0.57 \times \dfrac{120}{100}$ cumec

$= 0.68$ cumec

Example 67. Gross command area of an irrigation canal is 10,000 ha. Cultivable irrigated area is 60%. Irrigation is not required at *kharif*. During *rabi*, intensity of irrigation is 80%. What is the discharge required at the head of the canal, if the duty at its head 1,000 ha/cumec if time factor and capacity factor are 15/20 and 0.75, respectively.

Solution:

Cultivable irrigated area $\qquad = 10,000 \times \dfrac{60}{100}$

$\qquad\qquad\qquad\qquad\qquad\qquad = 6,000$ ha

Irrigated area in *rabi* $\qquad = 6,000 \times \dfrac{80}{100}$

$\qquad\qquad\qquad\qquad\qquad\qquad = 4,800$ ha

Discharge requirement at head $\quad = $ Area under irrigation/duty

$\qquad\qquad\qquad\qquad\qquad\qquad = \dfrac{4,800}{400}$

$\qquad\qquad\qquad\qquad\qquad\qquad = 12.0$ cumec

Given that, time factor $\qquad\qquad = \dfrac{15}{20}$

+. Full supply discharge at head end $= 12 \times \dfrac{20}{15}$ cumec

$\qquad\qquad\qquad\qquad\qquad\qquad = 16$ cumec

and design discharge $\qquad\qquad = 16 \times \dfrac{1}{0.75}$ cumec

$\qquad\qquad\qquad\qquad\qquad\qquad = 21.33$ cumec

Example 68. A water harvesting tank has to be designed for irrigating 40 ha area of a farm during *rabi* with canal system. The different crops to be grown, their area and irrigation requirements are given below. Find the capacity of the water harvesting tank, if canal loss is 20% and reservoir loss is 10%.

Crop	Base period (days)	Depth of irrigation (cm)	Area (ha)
Rapseed	30	10	20.0
Potato	40	15	10.0
Vegetables	120	30	4..0
Forage crops	90	15	5.0
Flower	120	30	1.0

Solution:

Crop	Duty (ha/cumec)	Discharge (cumec)	Volume of water required Cumec-days	Ha-m
	(1)	(2)	(3) = (2) × base period	(3) × 8.64
Rapseed	864 × 30/10 = 2592	20/2592 = 0.0077	0.2315	2.00
Potato	864 × 40/15 = 2304	10/2304 = 0.0043	0.1736	1.50
Vegetables	864 × 120/30 = 3456	4/3456 = 0.0012	0.1389	1.20
Forage crops	864 × 90/15 = 5184	5/5184 = 0.0001	0.0868	0.75
Flower	864 × 120/30 = 3456	1/3456 = 0.0003	0.0347	0.30
Total		**0.0136**	**0.6655**	**5.75**

Thus, total volume of water required by the crop $= 5.75$ ha-m

Volume of water required at head end $= 5.75 \times \dfrac{100}{80}$

$= 7.19$ ha-m

Water required to stored at the reservoir $= 16 \times \dfrac{100}{90}$

$= 7.99$ ha-m

Thus the capacity of the water storage tank should be $= 5.75$ ha-m

9.2.9. Water Potential

The term potential can be defined as the amount of work done or potential energy stored per unit mass in brining any mass 'm' from the reference level to the point in question.

+. Potential $= \dfrac{\text{Work}}{\text{Mass}}$

$= \dfrac{mgh}{m}$

$= gh$

If it is expressed in volume basis,

Potential $= \dfrac{\text{Work}}{\text{Volume}}$

$= \dfrac{mgh}{V}$

$= \dfrac{mgh}{m/r}$

$= \rho gh$

If it is expressed in weight basis,

Potential $= \dfrac{\text{Work}}{\text{Weight}}$

$= \dfrac{mgh}{mg}$

$= h$

Thus, potential is expressible physically in at least three ways:

1. Energy per unit mass: This is often taken to be fundamental expression of potential using units Erg/g or Joule/kg

   [1 bar = 1 × 10² Joules/kg = 1 × 10⁶ Ergs/g]

2. Energy per unit volume: Expression of energy per unit volume gives the dimension of pressure in dynes/cm², bar or atmosphere. This method of expression is more convenient for osmotic and pressure potential, but seldom used for gravitational potential.

3. Energy per unit weight (hydraulic head): Hydrostatic pressure can be expressed in terms of an equivalent hydraulic head, which is the height of liquid column corresponding to the given pressure.

    e.g., 1 atm = 1033 cm hydraulic head = 76 cm Hg

    When the soil water is at hydrostatic pressure greater than atmosphere, its pressure potential is considered as positive.

    When it is at pressure lower than atmospheric (a sub pressure commonly known as tension or suction), the pressure potential is considered as negative.

    Thus water under a free water surface is at zero pressure potential and water which has raisen in a capillary tube above that surface is characterized by a −ve pressure potential.

    The positive pressure potential which occurs below the ground water level has been termed as the "submerged potential"

    A negative pressure potential is termed "capillary potential" or "matric potential".

    Total soil water potential is defined as the amount of work that must be done per unit quantity of pure water in order to transport reversibly and isothermally an infinitesimal quantity of water from a pool of pure water at specific elevation at atmospheric pressure to the soil water (at the point under consideration).

    Water potential is given by –

    $$\Psi = \frac{RT}{V} \ln \frac{e}{e_0}$$

Where,

R = Gas constant (0.082 lit atm/degree-mole)

T = Absolute temperature

e = Actual vapour pressure

e_0 = Saturated vapour pressure

Example 69. Calculate the water potential of the atmosphere at (*a*) 99% (*b*) 90% and (*c*) 80% Relative Humidity at 27 °C.

Solution:

(*a*) *. Water potential, $\psi = \dfrac{RT}{V} \ln \dfrac{e}{e_0}$

+. Water potential at 99% RH $= \dfrac{0.082 \times 300}{0.018} \ln (0.99)$

[molal volume is not adjusted for temperature]

$= -13.74$ atm

$= -13.92$ bar

(*b*) Water potential at 90% RH $= \dfrac{0.082 \times 300}{0.018} \ln (0.90)$

$= -143.99$ atm

$= -145.86$ bar

(*c*) Water potential at 80% RH $= \dfrac{0.082 \times 300}{0.018} \ln (0.80)$

$= -304.96$ atm

$= -308.92$ bar

Water in soils and plants is subject to several fields caused by presence of the solid phase, dissolved salts, external gas pressure and the gravitational field. These effects are quantitatively expressed in terms of potential energy of water.

The soil and plant water potential (ψ) at any point in the system can be partitioned into-

(*i*) The osmotic potential ($\psi_{<!}$), due to presence of dissolved solutes.

(*ii*) The pressure potential (ψ_p) due to the turgor pressure acting out ward on the cell walls and internal membrane in plants, and in soil, it is related to the hydraulic or hydrostatic pressure found under saturated condition.

(iii) The matric potential (ψ_m) due to the forces of capillary and molecular imbibitional forces associated with the cell walls and colloidal surface which bind some of the water.

(iv) The gravitational potential (ψ_g) due to the gravity forces

Thus, $$\psi = \psi_{<!} + \psi_p \text{ or } \psi_m + \psi_g$$

Osmotic potential ($q_{<!}$) could be approximately estimated from the electrical conductivity of the soil solution at saturation:

$$\psi_{<!} = -0.36 \text{ EC}$$

Where,

$\psi_{<!}$ = osmotic potential in bar

EC = electrical conductivity in dS/m

Example 70. Calculate the total water potential of soil, if osmotic, matric and gravitational potential are 0, – 100 cm and – 50 cm of H_2O

Solution:

Given that,

$$\psi_{<!} = 0$$
$$\psi_m = -100 \text{ cm}$$
$$\psi_g = -50 \text{ cm}$$
$$\psi = 0 + (-100) + (-50)$$
$$= -150 \text{ cm of } H_2O$$

Example 71. Calculate the osmotic potential of soil solution, if EC of soil solution is 2 dS/m.

Solution:

Given that,

$$\text{EC} = 2 \text{ dS/m}$$
$$\psi_{<!} = -0.36 \text{ EC}$$
$$= -0.36 \times 2 \text{ bar}$$
$$= -0.72 \text{ bar}$$
$$= -0.72 \times 1020 \text{ cm of } H_2O$$
$$= -734.4 \text{ cm of } H_2O$$

Example 72. Calculate the osmotic potential of the soil solution at saturation, field capacity (FC), permanent wilting point (PWP) and at 20% soil moisture, if irrigation water having EC 5 dS/m is added to a salt free soil. Given that, volumetric moisture content at FC and PWP are 30% and 15%.

Solution:

*. $\psi_{<!}$ = – 0.36 EC

+. At saturation, $\psi_{<!}$ = – 0.36 × 5 bar

= – 1.80 bar

As water will freely flow from saturation to FC, the concentration of soil solution will not change.

So at FC,   EC = 5 dS/m

+. At saturation, $\psi_{<!}$ = – 0.36 × 5 bar

= – 1.80 bar

However, between FC and PWP, water will be lost due to transpiration and/or evaporation. Thus salt concentration will increase.

So, EC at 20% moisture   = (30/20) × 5 = 7.5 dS/m

and at 15% moisture   = (30/15) × 5 = 10 dS/m

+. $\psi_{<!}$ at 20% moisture   = – 0.36 × 7.5 bar

= – 2.70 bar

And at FC (15%) moisture = – 0.36 × 10 bar

= – 3.60 bar

Example 73. Water having EC 5 dS/m is added to a salt free soil of a drum to saturation. The bottom of the drum is closed and water is lost through evaporation only. Considering the bottom of the soil as reference level, calculate the total water potential 30 cm above the drum at volumetric moisture content 40% (saturation), 35%, 30%, 25%, 20% and 15%. The corresponding matric potentials at these moisture content were 0, – 0.005, – 0.03, – 1, – 5 and – 13 bar.

Solution:

Here, point 'A' is 30 cm above the reference level. Thus gravitational potential will be

$$= 30 \text{ cm of } H_2O$$

$$= 0.03 \text{ bar}$$

Thus total water potential at different moisture level:

Vol. Moisture content (%)	Ψ_{cl}(bar)	Ψ_m(bar)	Ψ_g(bar)	Ψ(bar)
40	$- 5 \times 0.36 = - 1.8$	0	0.03	$- 1.80 + 0 + 0.03 = - 1.770$
35	$- 5 \times (40/35) \times 0.36 = - 2.05$	$- 0.005$		$- 2.05 - 0.005 + 0.03 = - 2.025$
30	$- 5 \times (40/30) \times 0.36 = - 2.40$	$- 0.030$	0.03	$- 2.40 - 0.030 + 0.03 = - 2.400$
25	$- 5 \times (40/25) \times 0.36 = - 2.88$	$- 1.000$	0.03	$- 2.88 - 1.000 + 0.03 = - 3.850$
20	$- 5 \times (40/20) \times 0.36 = - 3.60$	$- 5.000$	0.03	$- 3.60 - 5.000 + 0.03 = - 8.570$
15	$- 5 \times (40/15) \times 0.36 = - 4.80$	$- 13.00$	0.03	$- 4.80 - 13.000 + 0.03 = - 17.77$

9.2.10. MATRIC POTENTIAL WITH THE HELP OF TENSIOMETER

Here,

d = depth of the tensiometer below soil surface

h_m = height of mercury column

h_{ms} = height of mercury level in the reservoir from the surface

$$y = d + h_m + h_{ms}$$

Thus, at B,

Gravitational potential, $\Psi_{gB} = h_m + h_{ms}$ [soil surface at reference level]

Matric potential, $\Psi_{mB} = - 13.6 \, h_m$ [–ve pressure of hanging column, thus represented by Ψ_m]

At D,

Gravitational potential, $\Psi_{gD} = - d$

Matric potential, $\Psi_{mD} = q_m$ (say)

Thus,   During equilibrium,

Total water potential at B = Total water potential at D

or $\Psi_{gB} + \Psi_{mB} = \Psi_{gD} + \Psi_{mD}$

or $h_m + h_{ms} - 13.6\ h_m = -d + \Psi_m$

or $\Psi = -13.6\ h_m + h_m + h_{ms} + d$

$= -13.6\ h_m + y$

Matric potential with manometer type tensiometer,
$\Psi_m = -13.6\ h_m + (h_m - h_{ms} + d)$

Example 74. A manometer type tensiometer was installed at 60 cm depth. Mercury container was kept at 10 cm above the surface. Calculate the matric potential and total water potential below 60 cm from the surface considering soil surface as reference level, if mercury rises to 25 cm.

Solution:

Here,

$$d = 60 \text{ cm}$$
$$h_m = 25 \text{ cm}$$
$$h_{ms} = 10 \text{ cm}$$

+.   Matric potential, $\Psi_m = -13.6 \times 25 + 25 + 10 + 60$

$$= -245 \text{ cm}$$

Gravitational potential, $\Psi_g = -60$ cm

Total water potential, $\Psi = \Psi_m + \Psi_g$

$$= -245 - 60$$
$$= -305 \text{ cm}$$

Example 75. Two manometer type tensiometers A and B were installed at 120 and 150 cm below the soil surface. Mercury containers were kept at 10 cm above the surface. If the mercury rise were 30 and 40 cm, calculate the total water potential at 120 cm and 150 cm below the soil surface and show the direction of flow.

Solution:

Matric potential of Tensiometer A, Ψ_{mA} = $-$ 13.6 × 25 + 25 + 10 + 60
$$= -248 \text{ cm}$$

Gravitational potential of Tensiometer A, Ψ_{gA} = $-$ 120 cm

Total water potential of Tensiometer A, Ψ_A = $-$ 248 cm $-$ 120 cm
$$= -368 \text{ cm}$$

Matric potential of Tensiometer B, Ψ_{mB} = $-$ 13.6 × 40 + 40 + 10 + 150
$$= -344 \text{ cm}$$

Gravitational potential of Tensiometer B, Ψ_{gB} = $-$ 150 cm

Total water potential of Tensiometer B, Ψ_B = $-$ 344 cm $-$ 150 cm
$$= -494 \text{ cm}$$

*. $\Psi_A > \Psi_B$, water will flow from A to B

Example 76. What will be the height of mercury column, if a manometric type tensiometer installed at 30 cm soil depth gives matric potential value of 0.1, 0.2, 0.3, 0.4, 0.5 and 0.6 atm? The level of mercury cup is 10 cm above soil surface.

Solution:

Given that, d = 60 cm

h_{ms} = 25 cm

*. $\Psi_m = -13.6\ h_m + h_m + h_{ms} + d$

+.   For 0.1 atm matric potential, $-0.1 \times 1033 = -12.6\ h_m + 40$

[since, 1 atm = 1033 cm]

or $h_m = (103.3 + 40)/12.6$
$$= 11.37 \text{cm}$$

Thus,

For 0.2 atmos matric potential, $h_m = (2 \times 103.3 + 40)/12.6 = 19.57$ cm

For 0.3 atmos matric potential, $h_m = (3 \times 103.3 + 40)/12.6 = 27.77$ cm

For 0.4 atmos matric potential, $h_m = (4 \times 103.3 + 40)/12.6 = 35.97$ cm

For 0.5 atmos matric potential, $h_m = (5 \times 103.3 + 40)/12.6 = 44.17$ cm

0For 0.6 atmos matric potential, $h_m = (6 \times 103.3 + 40)/12.6 = 52.37$ cm

[some students equilibrates height of mercury with 40 cm water height, which is not correct]

Example 77. Calculate the matric potential of soil, if a gauge type tensiometer with gauge factor 100 mm reads 60 and distance between gauge and tensiometer cup is 50 cm. The manufacturer has not calibrated for the length of the tensiometer.

Solution:

Matric potential with gauge type tensiometer, $\Psi_m = - F_g . R_g + Z_0$

Given that, F_g = 100 mm

                            = 10 cm

                 R_g = 60

                 Z_0 = 50 cm

$\therefore$ Matric potential, Ψ_m = $(- 10 \times 60 + 50)$ cm

                                 = $(- 600 + 50)$ cm

                                 = $- 550$ cm

Example 78. Calculate the matric potential of soil, if a gauge type tensiometer calibrated in bar reads 0.63 bar and distance between gauge and tensiometer cup is 50 cm. The manufacturer has not calibrated for the length of the tensiometer.

Solution:

Matric potential, Ψ_m = $(- 0.63 + 50/1020)$ bar [since, 1 bar = 1020 cm]

                         = $- 0.58$ bar

Example 79. Find out the matric potential of soil, if vacuum gauge tensiometer calibrated in inches with mercury reads 15 inch and distance between gauge and tensiometer cup is 50 cm. The manufacturer has not caliberated for the length of the tensiometer.

Solution:

Given that,   Reading   = 15 inch of Hg

                               = 15×13.6 inch of H_2O

                               = $15 \times 13.6 \times 2.54$ cm of H_2O

                               = 518.16 cm of H_2O

$\therefore$ Matric potential, Ψ_m = $(- 518.16 + 50.0)$ cm

                                 = $- 468.16$ cm

⋮ EXERCISES▶

1. If 120 cc soil contains 60 cc soil solids and 30 cc water, calculate the volume of pore (void), volume of air, void ratio, total porosity and degree of saturation.

2. If 100 g wet soil contains 20 g moisture, calculate the gravimetric moisture content.

3. If 60 cc soil contains 20 g moisture, calculate the volumetric moisture content.

4. Calculate the volumetric moisture content, if gravimetric moisture content and bulk density of soil is 15% and 1.4 g/cc.

5. Find out the bulk density, apparent specific gravity, gravimetric and volumetric moisture percent of soil from the following data:

   Mass of moisture box   = 40 g

   Mass of wet soil with moisture box   = 120 g

   Mass of oven dry *soil* with moisture box = 100 g

   Volume of soil   = 45 cc

6. If the mass of soil particles and their volume are 120 g and 45.3 cc, calculate the particle density of the soil.

7. If the particle density of soil minerals and density of organic matter or soil having organic matter content 0.90% are 2.6 g/cc and 1.30 g/cc, calculate the particle density of the soil.

8. Calculate the particle density of soil having following composition:

Soil constituents	Content	Particle density (g/cc)
Humus	0.82	1.3
Mineral A	39.18	2.6
Mineral B	25	2.5
Mineral C	20	2.7
Mineral D	15	2.8

9. Calculate the mass of dry soil of 1 ha furrow slice (15 cm) having bulk density 1.35 g/cc.

10. Calculate the total porosity of soil, if bulk density and particle density of the soil are 1.3 g/cc and 2.63 g/cc, respectively.

11. Calculate the total porosity of soil, if void ratio is 0.95.

12. Calculate the fraction of air content and degree of saturation, if volumetric moisture content and total porosity of soil are 25% and 48%.

13. Calculate the gravimetric and volumetric moisture content and degree of saturation from the following information:

    (*a*) Mass of wet soil = 230 g

    (*b*) Mass of dry soil   = 200 g

    (*c*) Bulk density   = 1.3 g/cc

    (*d*) Particle density   = 2.65 g/cc

14. Calculate the gravimetric and volumetric moisture content, bulk density, void ratio, porosity, degree or saturation and air filled porosity from the following information:

    (*a*) Mass of wet soil = 260 g

    (*b*) Mass of dry soil   = 230 g

    (*c*) Volume of soil   = 170 cc

    (*d*) Particle density   = 2.6 g/cc

15. Calculate the bulk density of soil from the following information:

    (*a*) Mass of dry soil   = 200 g

    (*b*) Mass of saturated soil = 270 g

    (*c*) Particle density   = 2.65 g/cc

16. Calculate the amount of water content in cm, in 0-20 cm soil profile, if the soil with bulk density 1.3 g/cc contains 18% gravimetric moisture.

17. Calculate the amount of water content in cm, in 0-90 cm soil profile from the following information:

Depth of soil layer (cm)	Moisture (%)	Bulk density (g/cc)
0-30	15.0	1.35
30-60	17.0	1.36
60-90	20.0	1.36

18. Calculate the water deficit of 0-20 cm soil layer in cm, if gravimetric moisture content and field capacity of the soil having bulk density 1.35 g/cc are 23.0% and 18%, respectively.

19. The maximum water holding capacity, field capacity and permanent wilting point for 0-30 cm soil depth are 47%, 23% and 10.5%. If the bulk density of the soil is 1.35 g/cc, calculate the (*a*) drainable water (*b*) available water and (*c*) unavailable water for 0-30 cm soil depth.

20. Find out the quantity of irrigation water to be applied to 60 cm soil depth with following values of moisture:

Depth of soil layer (cm)	Moisture (%)		Bulk density (g/cc)
	FC	Before irrigation	
0-20	24.0	15	1.35
20-40	24.5	16.5	1.35
40-60	23.8	18.1	1.37

21. Find out the quantity of irrigation water to be applied to 90 cm soil depth with following values of moisture:

Depth of soil layer (cm)	Moisture (%)		Bulk density (g/cc)
	FC	Before irrigation	
0-30	10.2 cm	16 %	1.38
30-60	24.0%	7.3 cm	1.39
60-90	10.5 cm	8.9 cm	1.40

22. If (*a*) 4 cm (*b*) 5 cm (*c*) 6 cm irrigation water is applied in the above exercise, how will water be distributed in 0-30, 30-60 and 60-90 cm soil profile?

23. Find out the wetting depth when 3.0 cm rainfall is received by a soil with following values of moisture (assume that there is no run off loss):

Depth of soil layer (cm)	Moisture (%)		Bulk density (g/cc)
	FC	Before irrigation	
0-20	24.0	18.3	1.40
20-40	24.5	20.0	1.39
40-60	23.8	21.6	1.41

24. The following data were obtained before irrigation. If 5 cm irrigation water is applied to 75 cm root zone depth, what is the effective depth of irrigation? How much water is wasted?

Depth of soil layer (cm)	Moisture (%)		Bulk density (g/cc)
	FC	Before irrigation	
0-25	24.0	20.4	1.40
25-50	24.5	22.1	1.39
50-75	23.8	23.0	1.41

25. If the net irrigation requirement is 4.0 cm and field efficiency is 75%, calculate the gross irrigation requirement in field.

26. Irrigation to upland rice is scheduled at 30% depletion of available soil moisture. Field capacity of soil in the effective root zone depth of 60 cm is 24.5% with a permanent wilting point of 10.5%. Which soil moisture content irrigation is to be scheduled at? Also calculate the depth of water, if bulk density of the soil is 1.35 g/cc.

27. Calculate the cumulative pan evaporation required for scheduling irrigation at IW/CPE ratio (*a*) 0.8 (*b*) 1.0 and (*c*) 1.2 with 5.0 cm irrigation water.

28. In a wheat crop. 6 cm irrigation was scheduled at IW/CPE ratio 1.0. If 2 cm effective rainfall comes before irrigation, calculate the CPE and depth of irrigation.

29. A 5,000 m^2 wheat plot is to be irrigated with 6 cm water through STW. If the discharge of the pump is 6.0 lit/hour. Calculate the pumping time, if field efficiency is 70%.

30. A 3.2 ha wheat field was irrigated at 40% depletion of available water for 6 hours to irrigate 1.2 m effective root zone of the crop. The soil moisture content at FC and PWP were 33s (v/v) and 15% (v/v), respectively. From the diversion box of the irrigation canal, water was released @150 litres/sec. However, only 120 litres/sec were delivered to the field. The run off loss in the field was 350 m^3. The depth of water penetrated measured at 10 points from head to the tail end were 1.2, 1.18, 1.15, 1.10, 1.05, 1.03, 1.00, 1.00, 0.95 and 0.92 m. Determine the water conveyance efficiency, water application efficiency, water storage efficiency and water distribution efficiency.

31. A pump is operated 8 hours a day and discharges 50,000 lit/hour. A paddy crop is irrigated with 5 cm depth of irrigation water at 11 days interval. Calculate the area commanded by water discharged by the pump. Assume that additional water required for land preparation is met by hired pump.

32. A pump having 80% application efficiency is operated 8 hours a day. An area of 15 ha with FC and PWP 33.6 cm/m and 15.7 cm/m, respectively is to be irrigated by the pump. Irrigation water is to be applied at 40% depletion of available soil moisture in the root zone. Losses in water conveyance are negligible. Calculate net depth of irrigation, design frequency, depth of water pumped per application and required discharge capacity of irrigation pump in ha-cm/day and litres/sec, if conveyance loss and peak rate of moisture use by the crop are 10% and 4.5 mm/day.

33. Calculate the Crop WUE and Field WUE of mustard, if productivity is 1100 kg/ha and ET and WR are 16.0 cm and 30.1 cm, respectively.

34. The moisture content of root zone of two nearby plot just before rainfall are 18 cm (irrigated plot) and 11 cm (rainfed plot) and the field capacity of root zone is 20 cm. Bunds are provided around the plots to check run off. Calculate the effective rainfall and deep

percolation in both the plots, if (*a*) 3 cm (*b*) 6 cm (*c*) 9 cm (*d*) 12 cm rainfall is received.

35. Calculate the ER between 1-15 May, 2013 from the following observation, if the storage capacity of root zone is 15.0 cm and initial soil moisture content at the root zone is 12.0 cm.

Date	Rainfall (cm)	ET (cm)
30.04.2013	-	-
01.05.2013	-	0.41
02.05.2013	-	0.42
03.05.2013	2.0	0.29
04.05.2013	-	0.33
05.05.2013	-	0.39
06.05.2013	3.0	0.30
07.05.2013	-	0.35
08.05.2013	-	0.38
09.05.2013	12.0	0.2
10.05.2013	-	0.25
11.05.2013	-	0.39
12.05.2013	-	0.42
13.05.2013	-	0.45
14.05.2013	3.0	0.31
15.05.2013	-	0.36

36. Total depth of water application for a crop of 125 days base period is 120 cm. what is the duty of water?

37. Calculate the delta of a crop with a base period of 115 days and 1,100 ha/cumec duty.

38. Gross command area of an irrigation canal is 20,000 ha. Cultivable irrigated area is 70%. Intensity of irrigation is 30 and 85% for *kharif* and *rabi*, respectively. What is the discharge required at the head of the canal, if the duty at its head is 500 and 1200 ha/cumec for *kharif* and *rabi*, respectively?

39. Cultivable command area of a canal is 1500 ha. Intensity of irrigation for sugarcane and *rabi* vegetables are 30% and 40% with duty 1,400 and 1,100 ha/cumec, respectively. Calculate the required discharge at head end if capacity factor is 0.85.

40. If peak demand is 1s% more than the average requirement in the above example, calculate the design discharge.

41. Water is released at the rate of 12 cumec at the head end. Calculate the area that can be irrigated, if duty at field is 120 ha/cumec and conveyance loss is 20%.

42. In a 1,000 ha canal command area, *kharif* rice - *toria* cropping sequence is followed. The intensity of *kharif* rice and *toria* are 90% and 30%, respectively. The duty of these crops at head end of water course are 800 and 1,200 ha/cumec. Find the discharge required, if peak demand is 20% more than the average requirement.

43. Gross command area of an irrigation canal is 20,000 ha. Cultivable irrigated area is 70%. Irrigation is not required at *kharif*. During *rabi,* intensity of irrigation is 70%. What is the discharge required at the head of the canal, if the duty at its head is 1,100 ha/cumec, time factor and capacity factor are 16/20 and 0.70, respectively?

44. A farmer decided to transplant early *ahu* rice in his farm. If base period and delta are 100 days and 80 cm, respectively, calculate the area the farmer can opt for cultivation of early *ahu* rice with the diesel pump available to him with n discharge of 40,000 lit/ha. Assume that water required for land preparation is 30% more than average requirement of the crop.

45. Calculate the water potential of the atmosphere at (*a*) 95% (*b*) 85% and (*c*) 75% relative humidity at 25°C.

46. Calculate the total water potential of soil, if osmotic, matric and gravitational potential are 0, -90 cm and -150 cm of H_2O.

47. Calculate the osmotic potential of soil solution, if EC of soil solution is 1.5 dS/m.

48. Calculate the osmotic potential of soil solution at saturation, field capacity (FC), permanent wilting point (PWP) and at 17% soil moisture, if irrigation water having EC 4 dS/m is added to a salt free soil. Given that, volumetric moisture content at FC and PWP are 33% and 16%.

49. A manometer type tensiometer was installed at 30 cm depth. Mercury container was kept at 10 cm above the surface. Calculate the matric potential and total water potential below 30 cm from the surface considering soil surface ns reference level, if mercury rises to 20 cm.

50. Two manometer type tensiometers A and B were installed at 90 and 100 cm below the soil surface. Mercury containers were kept at 10 cm above the surface. If the mercury rise were 30 and 25 cm, calculate the total water potential at 90 cm and 100 cm below the soil surface and show the direction of flow.

51. Calculate the matric potential of soil, if a gauge type tensiometer with gauge factor 100 mm reads 50 and distance between gauge and tensiometer cup is 60 cm. The manufacturer has not calibrated for the length of the tensiometer.

52. A gauge type tensiometer with a distance of 50 cm between gauge and tensiometer reads 0.4 bar. What will be reading in a tensiometer with a distance of 100 cm between gauge and tensiometer at the same point, if the manufacturer has not calibrated for the length of the tensiometers?

53. Calculate the matric potential of soil, if a gauge type tensiometer calibrated in bar reads 0.60 bar and distance between gauge and tensioneter cup is 50 cm. The manufacturer has not calibrated for the length of the tensiometer.

54. Find out the matric potential of soil, if vacuum gauge tensiometer calibrated in inches with mercury reads 16 inch and distance between gauge and tensiometer cup is 60 cm. The manufacturer has not calibrated for the length of the tensiometer.

⋮ ANSWERS ▶

1. Volume of pore = 60 cc; Volume of air = 30 cc; Void ratio = 1.0; Total porosity 50%; Degree of saturation = 50%

2. 25%   3. 33.3%   4. 21%

5. Bulk density = 1.33 g/cc; Apparent specific gravity = 1.33; Gravimetric moisture content = 33.33%; Volumetric moisture content = 44.44%

6. 2.65%   7. 2.59 g/cc   8. 2.61 g/cc

9. 2.25×10^6 kg   10. 50.57%   11. 48.72%

12. 52.08%

13. Gravimetric moisture content = 15%; Volumetric moisture content = 19.5%; Degree of saturation = 38.28%

14. Gravimetric moisture content = 13.04%; Volumetric moisture content = 17.64%; Bulk density = 1.35 g/cc; Void ratio = 0.92; Porosity = 47.96%; Degree of saturation = 36.79%; Air filled porosity = 30.31%

15. 1.37 g/cc   16. 4.68 cm   17. 23.17 cm

18. 1.35 cm

19. (a) 9.72 cm (b) 5.06 cm (c) 4.25 cm   20. 6.15 cm

21. 7.88 cm

22. (a) 3.58 cm (0-30 cm); 0.42 cm (30-60 cm); 0 (60-90 cm)

(b) 3.58 cm (0-30 cm); 1.42 cm (30-60 cm); 0 (60-90 cm)

(c) 3.58 cm (0-30 cm); 2.42 cm (30-60 cm); 0 (60-90 cm)

23. 44.94 cm

24. Effective irrigation = 2.38 cm; Wasted water = 2.62

25. 5.33 cm   26. 20.3%; 3.40 cm

27. (a) 6.25 cm (b) 5.0 cm (c) 4.17 cm

28. 6 cm; 4 cm   29. 71 hours 26 minutes

30. Water conveyance efficiency = 80%; Water application efficiency = 88.43%; Water storage efficiency = 82.90% and Water distribution efficiency = 92.48%.

31. 8.8 ha

32. Net depth of irrigation = 7.16 cm; Irrigation interval = 16 days; Depth of water pumped per application = 9.94 cm; Discharge capacity = 93.28 ha-cm/day or 32.29 lit/sec

33. Crop WUE = 68.75 kg/ha-cm; Field WUE = 36.54 kg/ha-cm

34. (*a*) Irrigated plot : ER = 2.0 cm; Deep percolation = 1.0 cm

Rainfed plot : ER = 3.0 cm; Deep percolation = 0

b) Irrigated plot : ER = 2.0 cm; Deep percolation = 4.0 cm

Rainfed plot : ER = 6.0 cm; Deep percolation = 0

(*c*) Irrigated plot : ER = 2.0 cm; Deep percolation = 7.0 cm

Rainfed plot : ER = 9.0 cm; Deep percolation = 0

(*d*) Irrigated plot : ER = 2.0 cm; Deep percolation = 10.0 cm

Rainfed plot : ER = 9.0 cm; Deep percolation = 3.0 cm

35. 7.58 cm   **36.** 908 ha/cumec   **37.** 90.33 cm

38. *Kharif* = 8.4 cumec; *Rabi* = 9.92 cumec   **39.** 1.02 cumec

40. 1.17 cumec   **41.** 1152 ha   **42.** 1.35 cumec

43. 15.91 cumec   **44.** 9.23 ha

45. (*a*) – 70.54 bar, (*b*) – 223.50 bar, (*c*) – 395.62 bar

46. – 240 cm   **47.** – 0.54 bar

48. At saturation = – 1.44 bar

At Field Capacity = –1.44 bar

At 17% Volumetric Moisture = – 2.80 bar

49. Matric potential = – 212 cm; Total water potential = – 242 cm

50. Total water potential at A = – 368 cm

Total water potential at B = – 305 cm

Water will move from B to A

51. – 440 cm   **52.** – 0.45 bar   **53.** – 0.55 bar

54. – .019 bar

CHAPTER 10

MOISTURE CONTENT IN PLANTS, GRAINS AND ORGANIC MANURES

The oldest method of measuring the water status of plants is in terms of water content as percentage of fresh weight or dry weight. But generally, moisture content of soil is expressed on dry weight basis (denominator is dry weight) and moisture content of plants, grains and organic manure is expressed in fresh weight basis (denominator is fresh weight). Since, moisture content determination in soil has already been discussed, water content determination of plants, grain and organic manures will not be discussed in this chapter.

10.1. MOISTURE CONTENT ON FRESH WEIGHT BASIS

Moisture content on fresh weight basis is determined by dividing moisture content of the material by fresh weight (initial weight) of the material and expressed in percentage.

$$\text{MC (\%)} = \frac{\text{Fresh weight - Dry weight}}{\text{Fresh weight}} \times 100$$

10.2. RELATIVE WATER CONTENT (RWC)

RWC is determined by dividing moisture content by saturated moisture content and expressed in percentage.

$$\text{RWC (\%)} = \frac{\text{Fresh weight - Dry weight}}{\text{Turgid weight - Dry weight}} \times 100$$

10.3. WATER DEFICIT OR SATURATION DEFICIT (SD)

It is calculated as –

$$\text{SD (\%)} = \frac{\text{Turgid weight - Fresh weight}}{\text{Turgid weight - Dry weight}} \times 100$$

$$= 1 - \frac{\text{Turgid weight - Fresh weight}}{\text{Turgid weight - Dry weight}} \times 100$$

$$= 100 - \frac{\text{Turgid weight - Fresh weight}}{\text{Turgid weight - Dry weight}} \times 100$$

$$= 100 - \text{RWC}$$

Example 1. Calculated moisture content, Relative Water Content (RWC) and Saturation Deficit (SD) from the following observations:

(*a*) Fresh weight of plant   = 50 g

(*b*) Oven dry weight of plant   = 5 g

(*c*) Turgid weight of plant   = 65 g

Solution:

$$\text{MC (\%)} = \frac{50 - 5}{50} \times 100$$

$$= 90$$

Hence moisture content of the plant is 90%

$$\text{RWC (\%)} \quad = \quad \frac{50 - 5}{65 - 5} \times 100$$

Hence RWC of plant is 75%

$$\text{SD (\%)} \quad = \quad \frac{50 - 5}{65 - 5} \times 100$$

Hence SD of plant is 25%

Example 2. Calculate the moisture content of a sample of vermicompost, if 200 g compost after oven drying weighs 140 g.

Solution :

$$\text{Moisture content of vermicompost} \quad = \frac{200 - 140}{200} \times 100\%$$

$$= \frac{60}{200} \times 100\%$$

$$= 30\ \%$$

Example 3. Calculate the moisture content in rice grain if 200 g rice grain yields 160 g after oven drying.

Solution :

$$\text{Moisture content of rice grain} \quad = \frac{200 - 160}{200} \times 100\% \times 100\ \%$$

$$= \frac{40}{200} \times 100\% \times 100\ \%$$

$$= 20\ \%$$

Example 4. If the weight of 4 ton rice grain just after harvest contains 20% moisture, calculate the weight at 14% moisture.

Solution:

$$20\% \text{ of 4 ton} \quad = 0.8 \text{ ton}$$

Dry matter $\qquad = (4 - 0.8)$ ton

$\qquad\qquad\qquad\qquad = 3.2$ ton

Thus, at 14% moisture, 86 ton dry matter will be present in = 100 ton grain

1 ton dry matter will be present in $\quad = (100/86)$ ton grain

3.2 ton dry matter will be present in $\; = 3.2 \times (100/6)$ ton grain

$\qquad\qquad\qquad\qquad\qquad\qquad\quad = 3.72$ ton grain

Thus, grain yield at 14% moisture $\quad = 3.72$ ton

Alternately,

Let, Mass of dry matter $\qquad\qquad\quad = M$

Mass of moisture at 20% moisture $\;\; = M_{20}$

Mass of moisture at 14% moisture $\;\; = M_{14}$

$$\frac{M_{20}}{M_{20} + M} = \frac{20}{100}$$

or $\qquad 1 - \dfrac{M_{20}}{M_{20} + M} = 1 - \dfrac{20}{100}$ [subtracting both side from 1]

or $\qquad \dfrac{M_{20}}{M_{20} + M} = \dfrac{100 - 20}{100}$ $\qquad\qquad$ ... (1)

and $\qquad \dfrac{M_{14}}{M_{14} + M} = \dfrac{14}{100}$

or $\qquad 1 - \dfrac{M_{14}}{M_{14} + M} = 1 - \dfrac{14}{100}$ [subtracting both side from 1]

or $\qquad \dfrac{M}{M_{14} + M} = \dfrac{100 - 14}{100}$ $\qquad\qquad$ ... (2)

223

Now, dividing equation (1) by equation (2)

$$\frac{M}{M_{14} + M} = \frac{100 - 14}{100}$$

or

$$M_{14} + M = \frac{100 - 20}{100 - 14} \times (M_{20} - M)$$

or   Grain yield at 14% moisture $= \dfrac{100 - 20}{100 - 14} \times$ Grain yield at 14% moisture

---

Grain yield at $\times$ % moisture $= \dfrac{100 - y}{100 - x} \times$ Grain yield at y % moisture

---

In the given example, grain yield at 20 % moisture = 40 ton

Grain yield at 14% moisture $= \dfrac{100 - 20}{100 - 14} \times 4$ ton

$$= 3.72 \text{ ton}$$

Example 5. If the weight of 4 ton rice grain just after harvest contains 19 % moisture, calculate the weight at 12 % moisture.

Solution

Grain yield at 12% moisture $= \dfrac{100 - 19}{100 - 12} \times$ Grain yield at 19% moisture

$$= \frac{81}{88} \times 4 \text{ ton}$$

$$= 3.68 \text{ ton}$$

⋮ EXERCISES ▶

1. Calculate moisture content, Relative Water Content (RWC) and Saturation Deficit (SD) from the following observations:

    (a) Fresh weight of plant   = 100 g

    (b) Oven dry weight of plant   = 11 g

    (c) Turgid weight of plant   = 128 g

2.   Calculate the moisture content of a sample of vermicompost, if 160 g compost after oven drying weighs 100 g.

3.   Calculate the moisture content in rice grain, if 150 g rice grain yields 115 g after oven drying.

4.   If the weight of 4 ton rice grain just after harvest contains 22% moisture, calculate the weight at 14% and 12% moisture.

⋮ ANSWERS ▶

1. Moisture content = 89%

    RWC   = 76.07%

    SD   = 23.93%

2. 37.5%

3. 23.33%

4. At 14% moisture = 3.63 ton

    At 12% moisture = 3.55 ton

GROWTH AND DEVELOPMENT ANALYSIS

Growth is irreversible increase in size and mass. Development is the phasic change. The study of growth and development helps us to understand the effect of different management practices as well as the performance of the cultivars. The different indices of growth and development are discussed in this chapter.

11.1 CROP GROWTH RATE (CGR)

It represents dry weight gained by a unit area of crop in a given time. It is expressed in $g/m^2/day$.

$$CGR = \frac{W_2 - W_1}{(t_2 - t_1)S}$$

Where,

W_1 and W_2 are crop plant dry weight (g) at time t_1 and t_2 respectively.

S is land area (m^2) over which dry matter was recorded.

Example 1: Plant dry weights of toria crop at 15,30 and 45 (days after sowing) are given in the following table. Calculate the CGR, if the spacing is 30 cm × 10 cm

Plant Dry Weight (g/Plant)		
15 DAS	**30 DAS**	**45 DAS**
0.27	2.16	4.71

Solution:

Area occupied by 1 plant= $0.3 \times 0.1 = 0.03$ m^2

Dry weight of 1 plant at 15 DAS=0.27 g

Dry weight of 1 plant at 30 DAS =2.16 g

Dry weight of 1 plant at 45 DAS=4.71 g

Therefore, CGR between 15- 30 DAS = $\dfrac{2.16 - 0.27}{(30-15) \times 0.03}$ = 4.20 g/m^2/day

Similarly, CGR between 30- 45 DAS = $\dfrac{4.71 - 2.16}{(45-30) \times 0.03}$ = 5.67 g/m^2/day

11.2 ABSOLUTE GROWTH RATE (AGR)

It expresses the dry weight increase per unit time and is expressed in g/plant/day.

$$AGR= \frac{W_2 - W_1}{(t_2 - t_1)}$$

Where,

W_1 and W_2 are crop plant dry weight (g) at time t_1 and t_2 respectively.

Example 2: Plant dry weights of toria crop at 15,30 and 45 (days after sowing) are given in the following table. Calculate the CGR, if the spacing is 30 cm × 10 cm

Plant Dry Weight (g/Plant)		
15 DAS	**30 DAS**	**45 DAS**
0.27	2.16	4.71

Solution:

Dry weight of 1 plant at 15 DAS=0.27 g

Dry weight of 1 plant at 30 DAS =2.16 g

Dry weight of 1 plant at 45 DAS=4.71 g

Therefore, AGR between 15- 30 DAS $= \dfrac{2.16 - 0.27}{(30\text{-}15)} = 0.13$ g/plant

Similarly, AGR between 30- 45 DAS $= \dfrac{4.71 - 2.16}{(45\text{-}30)} = 0.17$ g/plant

11.3 RELATIVE GROWTH RATE (RGR)

The relative growth rate of crops at time instant ® is defined as the increase of plant material per unit weight per unit time. It is expressed in g/g/day

$$RGR = \dfrac{\ln W_2 - \ln W_1}{(t_2 - t_1)}$$

Where,

W_1 and W_2 are crop plant dry weight (g) at time t_1 and t_2 respectively.

Example 3: Plant dry weights of toria crop at 15,30 and 45 (days after sowing) are given in the following table. Calculate the CGR, if the spacing is 30 cm × 10 cm

Plant Dry Weight (g/Plant)		
15 DAS	**30 DAS**	**45 DAS**
0.27	2.16	4.71

Solution:

Dry weight of 1 plant at 15 DAS = 0.27 g

Dry weight of 1 plant at 30 DAS =2.16 g

Dry weight of 1 plant at 45 DAS=4.71 g

Therefore, RGR between 15- 30 DAS $= \dfrac{\ln 2.16 - \ln 0.27}{(30\text{-}15)}$

$= \dfrac{0.7747 - (- 1.3093)}{(30\text{-}15)} = 0.139$ g/g/day

Similarly, RGR between 30- 45 DAS $= \dfrac{\ln 4.71 - \ln 2.16}{(45\text{-}30)}$

$= \dfrac{1.4279 - 0.7747}{(45\text{-}30)} = 0.0447$ g/g/day

11.4 LEAF AREA INDEX (LAI)

The leaf area is calculated by dividing the leaf area per plant by land area occupied by the plant

$$LAI= \dfrac{\text{Leaf Area}}{\text{Ground Area}}$$

Example 4: Leaf area of toria crop at 15,30 and 45 (days after sowing) are given in the following table. Calculate the LAI, if the spacing is 30 cm × 10 cm

Leaf Area (cm^2/Plant)		
15 DAS	**30 DAS**	**45 DAS**
45.37	361.84	475.82

Solution:

Area occupied by 1 plant = 30 cm × 10 cm= 300 cm^2

Leaf area of 1 plant at 15 DAS = 45.37 cm^2

Therefore, LAI at 15 DAS $= \dfrac{45.37}{300} = 0.15$

Similarly

LAI at 30 DAS $= \dfrac{361.84}{300} = 1.21$

LAI at 45 DAS $= \dfrac{475.82}{300} = 1.59$

11.5 LEAF WEIGHT RATIO (LWR) OR LEAF MASS FRACTION (LMF)

Leaf weight ratio is the mass of leaf to total dry mass of plant. It is a measure of allocation of leaf biomass.

$$LWR= \dfrac{\text{Mass of leaf}}{\text{Total mass of plant}}$$

Example 5: Mean plant dry weight and mean leaf dry weight of toria at 15,30 and 45 (days after sowing) are given in the following table. Calculate the LWR.

15 DAS		30 DAS		45 DAS	
Leaf dry weight (g/plant)	Total plant dry weight (g/plant)	Leaf dry weight (g/plant)	Total plant dry weight (g/plant)	Leaf dry weight (g/plant)	Total plant dry weight (g/plant)
0.16	0.27	1.22	2.16	1.67	4.71

Solution:

Leaf dry weight at 15 DAS=0.16 g

Total plant dry weight at 15 DAS= 0.27 g

Therefore, LWR$= \dfrac{0.16}{0.27} = 0.59$

Similarly,

Leaf dry weight at 30 DAS=1.22 g

Total plant dry weight at 30 DAS= 2.16 g

Therefore, LWR= $\dfrac{1.22}{2.16}$ = 0.56

Leaf dry weight at 45 DAS=1.67 g

Total plant dry weight at 45 DAS= 4.71 g

Therefore, LWR= $\dfrac{1.67}{4.71}$ = 0.35

11.6 LEAF AREA RATIO (LAR)

Leaf area ratio is the ratio of leaf area to total plant biomass. It is a measure of leafiness or photosynthesis surface relative to respiratory mass. It is expressed in cm^2/g.

$$LAR = \frac{\text{Leaf area of plant}}{\text{Total dry weight of plant}}$$

Example 6: Plant dry weight and leaf area of toria at 15,30 and 45 (days after sowing) are given in the following table. Calculate the LAR.

15 DAS		30 DAS		45 DAS	
Leaf area (cm^2/plant)	Total plant dry weight (g/plant)	Leaf area (cm^2/plant)	Total plant dry weight (g/plant)	Leaf area (cm^2/plant)	Total plant dry weight (g/plant)
47.37	0.27	361.84	2.16	475.82	4.71

Solution :

Leaf area at 15 DAS= 47.37 cm^2

Plant dry weight at 15 DAS= 0.27 g

Therefore, LAR= $\dfrac{47.37}{0.27}$ = 175.44 cm²/g

Similarly,

Leaf area at 30 DAS= 361.84 cm²

Plant dry weight at 15 DAS= 2.16 g

Therefore, LAR= $\dfrac{361.84}{2.16}$ = 167.52 cm²/g

Leaf area at 45 DAS= 475.82 cm²

Plant dry weight at 45 DAS= 4.71 g

Therefore, LAR= $\dfrac{475.82}{4.71}$ = 101.02 cm²/g

11.7 SPECIFIC LEAF AREA RATIO (SLA)

Specific leaf area is the ratio of leaf area to leaf mass. It is a measure of relative spread of leaf. It is expressed in cm²/g.

$$SLA= \dfrac{\text{Leaf area of plant}}{\text{Leaf dry weight}}$$

Example 7: Leaf dry weight and leaf area of toria at 15,30 and 45 (days after sowing) are given in the following table. Calculate the SLA.

15 DAS		30 DAS		45 DAS	
Leaf area (cm²/plant)	Total plant dry weight (g/plant)	Leaf area (cm²/plant)	Total plant dry weight (g/plant)	Leaf area (cm²/plant)	Total plant dry weight (g/plant)
47.37	0.16	361.84	1.22	475.82	1.67

Solution

Leaf area at 15 DAS= 47.37 cm²

Leaf dry weight at 15 DAS= 0.16 g

Therefore, SLA= $\dfrac{47.37}{0.16}$ = 296.06 cm²/g

Similarly,

Leaf area at 30 DAS= 361.84 cm²

Leaf dry weight at 15 DAS= 1.22 g

Therefore, SLA= $\dfrac{361.84}{1.22}$ = 296.59 cm²/g

Leaf area at 45 DAS= 475.82 cm²

Leaf dry weight at 15 DAS= 1.67 g

Therefore, SLA= $\dfrac{475.82}{1.67}$ = 284.92 cm²/g

11.8 SPECIFIC LEAF WEIGHT (SLW)

Specific leaf weight is the ratio of leaf dry weight to leaf area. It indicates the leaf thickness and density and is expressed as g/cm²

$$SLW= \frac{\text{Leaf dry weight}}{\text{Leaf area}}$$

Example 8: Leaf dry weight and leaf area of toria at 15,30 and 45 (days after sowing) are given in the following table. Calculate the SLA.

15 DAS		30 DAS		45 DAS	
Leaf area (cm²/plant)	Total plant dry weight (g/plant)	Leaf area (cm²/plant)	Total plant dry weight (g/plant)	Leaf area (cm²/plant)	Total plant dry weight (g/plant)
47.37	0.16	361.84	1.22	475.82	1.67

Solution

Leaf area at 15 DAS= 47.37 cm^2

Leaf dry weight at 15 DAS= 0.16 g

Therefore, SLW= $\dfrac{0.16}{47.37}$ = 0.0034 g/cm^2

Similarly,

Leaf area at 30 DAS= 361.84 cm^2

Leaf dry weight at 15 DAS= 1.22 g

SLW = $\dfrac{1.22}{361.84}$ = 0.0034 g/cm^2

Leaf area at 45 DAS= 475.82 cm^2

Leaf dry weight at 15 DAS= 1.67 g

SLW= $\dfrac{1.67}{475.82}$ = 0.0035 g/cm^2

11.9 LEAF AREA DURATION (LAD)

Leaf area duration is the integral of leaf area index over a growth period and is expressed in days. Leaf area duration of a crop is measure of its ability to produce leaf area on unit area of land over a time period.

$$LAD = \frac{LAI_1 + LAI_2 \times (t_2 - t_1)}{2}$$

Where,

LAD = Leaf Area Duration between time t_2 and t_1

LAI_1 = Leaf Area Index at time t_1

LAI_2 = Leaf Area Index at time t_2

Example 9: Leaf area index of toria at 15,30,45, 60 and 75 (days after sowing) are given in the following table. Calculate the LAD.

Leaf Area Index				
15 DAS	30 DAS	45 DAS	60 DAS	75 DAS
0.15	1.21	1.59	0.51	0.03

Solution:

LAD between 0-15 DAS= $\dfrac{0+0.15}{2} \times (15-0) = 1.13$ days

LAD between 15 -30 DAS= $\dfrac{0.15 + 1.21}{2} \times (30-15) = 10.20$ days

LAD between 30 -45 DAS= $\dfrac{1.21 + 1.59}{2} \times (45-30) = 21.00$ days

LAD between 45-60 DAS= $\dfrac{1.59 + 0.51}{2} \times (60-45) = 15.75$ days

LAD between 60-75 DAS= $\dfrac{0.51 + 0.03}{2} \times (75-60) = 4.05$ days

LAD between 0 -75 DAS= 1.13 +21.00 +15.75 +4.05 = 52.13 days

11.10 BIOMASS DURATION (BMD)

It is integral of total biomass over a growth period and expressed in g days.

$$BMD = \frac{TDM_1 + TDM_2 \times (t_2 - t_1)}{2}$$

Where,

BMD = BioMass Duration between time t_2 and t_1

TDM_1 = Total Dry Matter at time t_1

TDM_2 = Total Dry Matter at time t_2

Example 10 : Plant dry weight of toria crop at 15,30 and $5 DAS are given in the following table. Calculate BMD.

Plant dry weight (g/plant)		
15 DAS	**30 DAS**	**45 DAS**
0.27	2.16	4.71

Solution:

BMD between 0-15 DAS= $\dfrac{0 + 0.27}{2} \times (15 - 0) = 2.03$ g days

BMD between 15-30 DAS= $\dfrac{0.27 + 2.16}{2} \times (30\text{-}15) = 17.48$ g days

BMD between 30-45 DAS= $\dfrac{2.16 + 4.71}{2} \times (45\text{-}30) = 51.53$ g days

BMD between 0-45 DAS= 2.03 + 17.48 + 51.53 = 71.04 g days

11.11 NET ASSIMILATION RATE

It indirectly indicates the rate of net photosynthesis. It is expressed as gram of dry matter production per day per m^2 leaf area.

$$NAR = \frac{(W_2 - W_1) \times (\ln L_2 - \ln L_1)}{(t_2 - t_1) \times (L_2 - L_1)}$$

Where,

NAR= Net Assimilation Rate

L_1 & W_1 = Leaf area and dry weight of the plant , respectively at time t_2

L_2 & W_2 = Leaf area and dry weight of the plant , respectively at time t_2

Example 11: Leaf area, leaf dry weight and plant dry weight of toria at 15,30 and 45 DAS are given in the following table. Calculate NAR.

Parameters	15 DAS	30 DAS	45 DAS
Leaf area (cm^2/plant)	45.37	361.84	475.82
Leaf dry weight (g/plant)	0.16	1.22	1.67
Leaf dry weight (g/plant)	0.27	2.16	4.71

Solution:

$$\text{NAR between 15-30 DAS} = \frac{(2.16\text{-}0.27) \times (\ln 361.84 - \ln 45.37)}{(30\text{-}15\,) \times (361.84\text{-}45.37)}$$

$$= \frac{1.89 \times 2.0735}{15 \times 316.47} = 8.25 \text{ g/day/ m}^2$$

$$\text{NAR between 30-45 DAS} = \frac{(4.71\text{-}2.16) \times (\ln 475.82 - \ln 361.84\,)}{(30\text{-}15\,) \times (361.84\text{-}45.37)}$$

$$= \frac{2.55 \times 0.2738}{15 \times 113.98} = 4.08 \text{ g/day/ m}^2$$

11.12 HARVEST INDEX

It is calculated by dividing teconomic yield by biological yield. It is expressed in percentage

$$\text{HI} = \frac{\text{Economic yield}}{\text{Biological yield}} \times 100$$

Example 12: Calculate the harvest index of rice from the following information:

(a) Grain yield = 4.3 t/ha

(b) Straw yield = 6.5 t/ha

Solution:

Here, Economic yield= 4.3 t/ha

Straw yield = 6.5 t/ha

Therefore, biological yield = (4.3 + 6.5) = 10.8 t/ha

$$\text{Harvest Index (\%)} = \frac{\text{Economic yield}}{\text{Biological yield}} \times 100$$

$$= \frac{4.3}{10.8} \times 100 = 39.8 \ \%$$

11.13 LEAF PRODUCTION RATE (LPR)

It can be estimated by counting the number of leaves on tagged plants at periodic intervals. It is expressed as number of leaves/day.

$$LPR = \frac{L_{n2} - L_{n1}}{(t_2 - t_1)}$$

Where,

L_{n1} = Number of leaves at times t_1

L_{n2} = Number of leaves at times t_2

Example 13: If the mean leaf number of maize at 25 and 40 DAS are 5 and 10, calculate the leaf production rate.

Solution:

LPR of maize between 25-40 DAS $= \dfrac{10 - 5}{40 - 25} = \dfrac{5}{15} = 0.33$ leaf/day

11.14 TILLER PRODUCTION RATE (TPR)

It can be estimated by counting the number of tillers on tagged plants at periodic intervals. It is expressed as number of tillers/day.

$$TPR = \frac{T_{n2} - T_{n1}}{(t_2 - t_1)}$$

Where,

T_{n1} = Number of tillers at times t_1

T_{n2} = Number of tillers at times t_2

Example 14: If the mean leaf number of tillers of rice crop at 30 and 50 DAS are 10/hill and 30/hill, calculate the tiller production rate.

Solution:

TPR of maize between 30-50 DAS $= \dfrac{30-10}{50-30} = \dfrac{20}{20} = 1.0$ tiller/day/hill

11.15 FLOWERING RATE (FR)

It indicates the synchronousness of flowering. It is expressed in number of flowers appears/day.

$$FR = \frac{F_{r2} - F_{r1}}{(t_2 - t_1)}$$

Where,

F_{r1} = Number of flowers that appear per plant at times t_1

F_{r2} = Number of flowers that appear per plant at times t_2

Example 15: If the first flower of gladiolus flowered 5 days after emergence of spike and 7th flower opened on 8 days after emergence of flower, calculate the flowering rate.

Solution:

FR of gladiolus $= \dfrac{7-1}{8-5} = \dfrac{6}{3} = 2$ flowers/day

11.16 PANICLE DENSITY (PD)

It indicates the compactness of panicle or earhead. It is calculated as

$$PD = \frac{\text{Number of grains or weight of grains/panicle}}{\text{Length of panicle}}$$

Example 16: If a rice variety PUSA-1121 having panicle length 25 cm contains 255 seeds, calculate the density of panicle

Solution:

$$PD = \frac{255}{25} = 10.2 \text{ seeds/cm}$$

⋮ EXERCISES ▶

1. Plant dry weight of toria at 15,30 and 45 DAS are given in the following table:

Plant dry weight (g/plant)		
15 DAS	**30 DAS**	**45 DAS**
0.30	2.30	5.1

If the spacing is 30 cm × 10 cm, calculate

   (a) CGR between 15-30 DAS and 30-45 DAS

   (b) RGR between 15-30 DAS and 30-45 DAS

   (c) AGR between 15-30 DAS and 30-45 DAS

   (d) BMD between 15-30 DAS and 30-45 DAS

2. Leaf area of toria at 15, 30 and 45 DAS are given in the following table. Calculate the LAI, if the spacing is 30 cm × 10 cm.

Leaf area (cm^2/plant)		
15 DAS	**30 DAS**	**45 DAS**
50.3	370.5	480.1

3. Mean plant dry weight and mean leaf dry weight of toria at 15 ,30 and 45 DAS are given in the following table. Calculate LWR.

15 DAS		30 DAS		45 DAS	
Leaf dry weight (g/plant)	Total plant dry weight (g/plant)	Leaf dry weight (g/plant)	Total plant dry weight (g/plant)	Leaf dry weight (g/plant)	Total plant dry weight (g/plant)
0.15	0.26	1.25	2.20	1.60	4.63

4. Plant dry weight and leaf area of toria at 15 ,30 and 45 DAS are given in the following table. Calculate LAR.

15 DAS		30 DAS		45 DAS	
Leaf area (cm^2/plant)	Total plant dry weight (g/plant)	Leaf area (cm^2/plant)	Total plant dry weight (g/plant)	Leaf area (cm^2/plant)	Total plant dry weight (g/plant)
47.3	0.30	369.3	2.21	482.3	4.83

5. Leaf area, leaf dry weight and plant dry weight of toria at 15 ,30 and 45 DAS are given in the following table. The plants are spaced at 30 cm × 10 cm.

15 DAS			30 DAS			45 DAS		
Leaf area (cm^2/plant)	Leaf dry weight (g/plant)	Total plant dry weight (g/plant)	Leaf area (cm^2/plant)	Leaf dry weight (g/plant)	Total plant dry weight (g/plant)	Leaf area (cm^2/plant)	Leaf dry weight (g/plant)	Total plant dry weight (g/plant)
48.3	0.18	0.30	373.2	1.25	2023	483.2	1.72	4.92

Calculate

(a) SLA at 15,30 and 45 DAS

(b) SLW at 15,30 and 45 DAS

(c) LAI at 15,30 and 45 DAS

(d) LAD between 15,30 and 45 DAS

(e) NAR between 15,30 and 45 DAS

6. Calculate the harvest index of rapeseed from the following observations:

    (a) Seed yield = 955 kg/ha

    (b) Stover yield= 2050 kg/ha

7. If the mean leaf number of maize at 25 and 40 DAS are 6 and 10, calculate the leaf production rate

8. If the mean number of tillers at 30 and 50 DAS are 11/hill and 35/hill. Calculate tiller production rate.

9. If the first flower of the gladiolus flowerd 5 days after emergence of spike and 8th flower opened on 8 days after emergence of flower. Calculate flowering rate.

10. If a rice variety having panicle length 22 cm contains 205 seeds, calculate the panicle density.

ANSWERS

1. (a) CGR= 4.44 g/m^2/day (15-30 DAS); 6.22 g/m^2/day (30-45 DAS)

    (b) RGR= 0.136g/g/day (15-30 DAS); 0.053g/g/day (30 -45 DAS)

    (c) AGR= 0.133 g/plant/day(15-30 DAS); 0.187 g/plant/day(30-45 DAS)

    (d) BMD =2.25 g.days(0-15DAS); 19.5 g.days(15-30 DAS); 55.5 g.days(30-45 DAS)

2. 0.17 (15 DAS);1.24 (30 DAS);1.60 (45 DAS)

3. 0.58 (15 DAS);0.57 (30 DAS);0.35 (45 DAS)

4. 157.7 cm^2/g (15 DAS);167.1 cm^2/g (30 DAS);99.86 cm^2/g (45 DAS)

5. (a) SLA=268.3 cm^2/g (15 DAS);298.6 cm^2/g (30 DAS);280.9 cm^2/g (45 DAS)

    (b) SLW= 0.0037 g/ cm^2(15 DAS);0.0033 g/cm^2 (30 DAS);0.0036 g/ cm^2 (45 DAS)

    (c) LAI= 0.16 (15 DAS);1.24 (30 DAS);1.61 (45 DAS)

(d) LAD =1.2 days(0-15DAS); 10.5 days(15-30 DAS); 21.4 days(30-45 DAS)

(e) 8.10 g/day/m²(15-30 DAS); 4.21 g/day/m² (30-45 DAS)

6. 31.78

7. 0.27 leaf/day

8. 1.2 tillers/day/hill

9. 2.33 flowers/day

10. 9.32 seeds/cm

ECONOMIC STUDIES

E conomic study is an important component of all agronomic experiments. When a new technology is developed, its economic viability should be tested. The commonly used economic indices are discussed in this chapter:

12.1 NET RETURN

Net return is calculated as

Net return= Gross return- Cost of Cultivation

Example 1: If gross returns and cost of cultivation of 1 hectare is Rs. 52,000 and Rs. 24,500/- respectively, calculate net returns/ha

Solution :

Net Return= Rs. 52,000- Rs. 24,500 = Rs. 27,500

12.2 RETURN PER RUPEE INVESTED

It is calculated as

$$\text{Return per Rupee Invested} = \frac{\text{Gross returns}}{\text{Cost of Cultivation}}$$

Example 2: If gross return and cost of cultivation of 1 hectare is Rs. 52,000 and Rs. 24,500/- respectively, calculate return per rupee invested

Solution :

Gross returns: Rs. 52,000

Cost of cultivation: Rs.24,500

$$\text{Return per Rupee Invested} = \frac{52,000}{24,500} = 2.12$$

12.3 BENEFIT COST RATIO

It is calculated as

$$\textbf{B: C ratio} = \frac{\text{Net returns}}{\text{Cost of Cultivation}}$$

Example 3: If gross returns and cost of cultivation of 1 hectare is Rs. 52,000 and Rs. 24,500/- respectively, calculate B: C ratio

Solution:

Net returns= Rs.52,000 − Rs. 24,500= Rs.27,500

Cost of Cultivation= Rs.24,500

$$\text{B: C ratio} = \frac{\text{Net returns}}{\text{Cost of Cultivation}} = 1.12$$

12.4 MARGINAL BENEFIT COST RATIO

The suitability of new technology could be assessed by determining marginal benefit cost ratio.

$$\text{MBCR} = \frac{\text{Gross return of new technology} - \text{Gross return of farmer's practice or Existing technology}}{\text{Total variable cost of new technology} - \text{Total variable cost of farmer's practice or Existing technology}}$$

Example 4: Calculate MBCR. The gross returns and cost of surface drain of sesamum are given below:

Treatment	Total Variable Cost (Rs./ha)	Gross returns (Rs./ha)
Farmer's practice	9,885	31,680
Surface drain	10,885	41,160

Solution:

$$\text{MBCR} = \frac{41,160.00 - 31680.00}{10,885.00 - 9885.00} = 9.48$$

12.5 RELATIVE ECONOMIC EFFICIENCY

The relative economic efficiency of new technology could be assessed by determining relative economic efficiency.

REE

$$= \frac{\text{Net returns of the improved system - Net returns of the existing system}}{\text{Net returns of the existing system}}$$

Example 4: Calculate REE. The gross returns and cost of surface drainage of sesamum are given below:

Treatment	Total Variable Cost (Rs./ha)	Gross returns (Rs./ha)
Farmer's practice	9,885	31,680
Surface drain	10,885	41,160

Solution:

Net returns of the surface drain= Rs. 41,600- Rs.10,885= 30,275.00

Net returns of the farmer's practice= Rs.31680- Rs. 9885 = 21,795

$$\text{REE} = \frac{30,275 - 21,795}{21795} = 0.39$$

12.6 PARTIAL FACTOR PRODUCTIVITY

It is the ratio of total output to a single output/ labour, output/machine, output/capital or output/energy. The output and input of PFP can be expressed in the same unit of measure or output per rupee invested in the input.

12.7 MULTI FACTOR PRODUCTIVITY

It utilizes more than a single factor. For example both labor and capital. Hence multifactor productivity is the ratio of total output to a subset of inputs. A subset of inputs might consist of only lsabour and material or it could include capital. The different factors must be measured in the same units. For example rupee or standard hours.

12.8 TOTAL FACTOR PRODUCTIVITY

It is measured by combining the effects of all the resources used in the production of good and services (labour,capital,raw material, energy etc.) and dividing into the output. Total output (grain and straw) must be expressed in the same unit of measure. Similarly, total input (labour, fertilizer, seed cost, herbicide cost etc.) must be expressed in the same unit of measure. However, the ratio of total output and total input need not to be expressed in the same unit of measure. Resources are often converted to rupee or standard hours to that a single figure can be used as an aggregate measure of total input or output. For example, total output could be expressed as weight of unit produced (e.g. grain yield of rice) or the number of units produced (number of flower produced) and total input could be expressed in rupees such as rice grain per rupee input. Other varieties of the measure may appear as rupee value of good or service produced per rupee of input or standard hours of output per actual hours of input.

Example 5: If the grain yield of rice with 0,20,40 and 60 kg N /ha are 24,32,40 and 43 q/ha and cost of 1 kg urea is Rs. 12.00. Calculate PFP of N.

Solution:

Cost of 20 kg N= 20 × 2.17 × 12.= Rs.520.80

Cost of 40 kg N= 40 × 2.17 × 12.= Rs.1041.60

Cost of 60 kg N= 60 × 2.17 × 12.= Rs.1562.40

Nitrogen (kg/ha)	PFP (q rice/kg N)	PFP (kg rice/kg N)
20	$\frac{32}{20}$ = 1.620	$\frac{3200}{520.8}$ = 6.14520.8
40	$\frac{40}{40}$ = 1.040	$\frac{4000}{10416}$ = 3.8410416
60	$\frac{43}{60}$ = 0.760	$\frac{4300}{1562.4}$ = 2.751562.4

Example 6: The yield of rice and inputs consumed for production of rice are given below:

(a) Yield of paddy: 5000 kg/ha

(b) Price of paddy: Rs.10,000/kg

(c) Cost of land preparation: Rs. 3,000

(d) Labour: Rs.15,000

(e) Fertilizer: Rs. 3,000

(f) Herbicides: Rs.1,000

(g) Miscellaneous: Rs. 2,000

Calculate

(a) PFP of labour, fertilizer and herbicides

(b) Calculate MFP of labour+ fertilizer and labour + fertilizer+herbicide

(c) Calculate TFP of rice culktivation

Solution:

Particulars	Factors	PFP(kg rice/rupee invested in labour, fertilizer or herbicides	PFP(Rs./Re. invested in labour, fertilizer or herbicides
(a)	Labour	$\frac{5000}{15000} = 0.33$	$\frac{50,000}{15000} = 3.33$
	Fertilizer	$\frac{5000}{3000} = 1.67$	$\frac{50,000}{3000} = 1.67$
	Herbicides	$\frac{5000}{19000} = 5.00$	$\frac{50,000}{1000} = 50.0$
(b)	Factors	MFP (kg rice/rupee invested in labour, fertilizer or herbicides	MFP(Rs./Re. invested in labour, fertilizer or herbicides
	Labour+ **Fertilizer +** **Pesticides**	$\frac{5000}{19000} = 0.261$	$\frac{50,000}{19000} = 2.63$
(c)	**TFP(kg rice/rupee invested)**		**TFP(Rs./Re. invested)**
	$\frac{50,000}{24000} = 0.21$		$\frac{50,000}{24000} = 2.08$

EXERCISES

1. If gross returns and cost of cultivation of 1 ha. rapeseed is Rs.54,000/- and Rs.18,000/-, respectively. Calculate the net returns/ha, return per rupee invested and B: C ratio.

2. The gross returns and cost of variable cost in rapeseed cultivation are given below:

Treatment	Total variable cost (Rs./ha)	Gross returns (Rs./ha)
Farmer's practice	16,800.00	33,000.00
Two irrigations at flowering and pod formation	18,000.00	54,000.00

Calculate MBCR and REE.

3. The yield of rapeseed and inputs consumed for the production of rapeseed are given below:

    (a) Yield of rapessed: 900 kg/ha

    (b) Price of rapeseed: Rs.30.00/kg

    (c) Cost of land preparation: Rs.3,000

    (d) Labour: Rs.10,000.00

    (e) Fertilizer: Rs.4,000.00

    (f) Herbicides: Rs.500.00

    (g) Miscellaneous: Rs.15,00.00

⋮ ANSWERS ▶

1. Net returns = Rs.36,000; Return per rupee invested = 3.00; B: C ratio= 2:1

2. **MBCR= 17.5; REE= 1.22**

3.

Particulars	Factors	PFP(kg rice/rupee invested in labour, fertilizer or herbicides	PFP(Rs./Re. invested in labour, fertilizer or herbicides
(a)	Labour	0.090	2.70
	Fertilizer	0.225	6.75
	Herbicides	1.800	54.00
(b)	Factors	MFP(kg rice/rupee invested in labour, fertilizer or herbicides	MFP(Rs./Re. invested in

			labour, fertilizer or herbicides
	Labour+ Fertilizer+ Pesticides	0.062	1.86
(c)	TFP(kg rice/rupee invested)		TFP(Rs./Re. invested)
	0.047		1.42

For Product Safety Concerns and Information please contact our EU
representative GPSR@taylorandfrancis.com
Taylor & Francis Verlag GmbH, Kaufingerstraße 24, 80331 München, Germany